OHIXIHO

A Biography of Charles Clark Fraser

Winston C. Fraser

© 2016 Winston Fraser Consulting Inc.

216 rue de l'Église
Rosemere, QC J7A 2X2
(450) 621-2378
wcfraser@sympatico.ca

All rights reserved. No part of this book may be adapted, reproduced or transmitted in any form or by any means, electronic, mechanical, photocopying, recording, microrecording, or otherwise, without the written permission of Winston Fraser Consulting Inc.

Layout and production: Jim Fraser

Front cover photo: Courtesy of Charles W. K. Fraser

Printed and bound in Canada by:

Katari Imaging
282 Elgin St.
Ottawa, ON K2P 1M3
613-233-1999

ISBN: 978-0-9950842-0-9

Contents

Dedication 4

Preface 5

Acknowledgements 6

Chapter 1   Charlie the Boy 7

Chapter 2   Charlie the Prospector 13

Chapter 3   Charlie the Astronomer 23

Chapter 4   Charlie the Inventor 31

Chapter 5   Charlie the Mathematician 41

Chapter 6   Charlie the Athlete 49

Chapter 7   Charlie the Fitness Fanatic 63

Chapter 8   Charlie the Chemist 73

Chapter 9   Charlie the Collector 93

    Numismatist 93

    Philatelist 95

    Curios Collector 103

Chapter 10   Charlie the Jack of Many Trades 109

    Confectionery Manufacturer 111

    Fox Rancher 109

    Woodsman 113

    Animal Trainer 115

    Photographer 116

    Carpenter 118

Chapter 11   Charlie the Philanthropist 119

Chapter 12   Charlie the Man 125

Epilogue 141

Appendix A   Autobiographical Summary 143

Appendix B   Fraser's Halo Complex of the Sun – March 21, 1930 ... 145

Appendix C   Birth/Baptism Record, Death Notice and Obituary 149

Appendix D   Frasier, Thornton & Company Raw Materials Price List 151

Dedication

This book is lovingly dedicated to my darling wife of 48 years, Becky, who passed away very suddenly just prior to the production phase. An admirer of Charlie ever since they first met, she enjoyed our frequent visits to his home during his later years, and was very proud to name our son Charles after him.

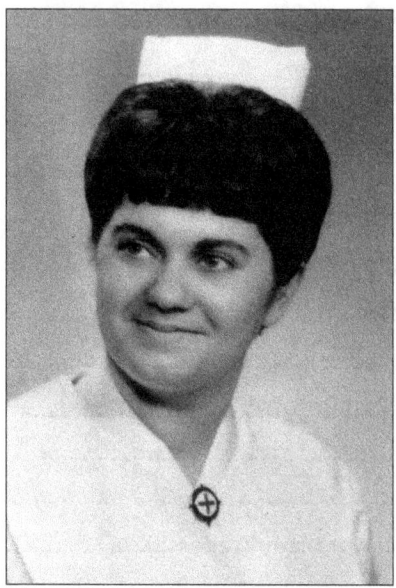

Preface

This book represents the fulfilment of a request made by my dad's first cousin, Charlie Fraser, in May 1973, when, at the age of 92, he gave me a somewhat weathered brown envelope stuffed with documents to be used in writing his official biography "later." It is unclear what he meant by "later," but now, after 43 years, his request is finally being honoured. The outside of the brown envelope is reproduced below.

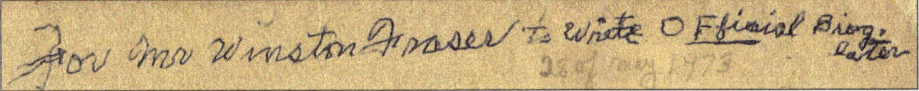

Had he been born a few generations later, Charlie might have become another Steve Jobs (co-founder of Apple Computer) or Usain Bolt (world champion sprinter). Born in 1881 in the small town of Cookshire in Quebec's Eastern Townships, Charles Clark Fraser was a man who was, in many respects, miles ahead of his time. In spite of significant disabilities, this remarkable man's long life was punctuated by many unique accomplishments in a wide variety of fields. Rarely has a single individual practiced so many vocations over a lifetime. To borrow a line from the 18th century nursery rhyme "Rub-a-dub-dub," he was neither "a butcher, a baker nor candlestick maker," but he was almost everything else. Charlie was a gold prospector, an inventor, a woodsman, an athlete, a mathematician, a chemist, a fox rancher, a numismatist, a philatelist, a photographer, a museum curator, an astronomer, an animal trainer, a dentist, a philanthropist and a jack of all trades. Although a lifelong bachelor, he was, strangely enough, very much a family man.

While known to most people simply as "Charlie," he used different monikers when referring to himself. In correspondence and legal documents he signed his name as "C.C. Fraser." In less formal notes he used only his initials, "CCF." Very often his signature was accompanied by the word "OHIXIHO" as though it was his address or even a part of his name. The real meaning of OHIXIHO is left for the reader to discover.

Acknowledgements

I wish to acknowledge the kind assistance provided to me during the research phase by members of Charlie's family as well as by friends and associates. In particular, I want to thank Charlie's niece Gloria (Frasier) Bellam and grandnephews Frasier, Liles and Harry Bellam; California cousins Sally Aldinger and the late Sharon Peart Waite; Canadian cousins Charles W. K. Fraser, Donald Parsons and my "Fraser 12" siblings; Cookshire friends Dorothy Ross and Patricia Stevenson Smith; and Maplemount Home associates Dr. Robert Paulette and Ruby Jean Campbell.

I would be remiss to not mention the valuable help of Sharon Moore, president of the Compton County Historical and Museum Society; genealogists Paul Lessard and Marie Fraser of the Clan Fraser Society of Canada; sketch artist James Harvey; and digital photo retoucher Greg Beck. Finally, a special thank you to my brother Jim for his expert proofreading, layout and production services as well as for his advice, support and encouragement throughout the project.

– Winston C. Fraser

Chapter 1   Charlie the Boy

His humble beginnings belied a life of remarkable accomplishments. Charles Clark Fraser was born in Cookshire, Quebec, Canada on June 29, 1881, the seventh of eight siblings:
- Bailey Elkins Frasier (1870-1876)
- Lilly Gertrude Frasier (1871-1871)
- Jared Cook Fraser (1873-1952)
- James Andrew Frasier (1875-1959)
- Ellen Amelia Frasier (1877-1957)
- Henry Rankin Frasier (1879-1948)
- Charles Clark Fraser (1881-1978)
- Hattie Fanny Maria Frasier (1886-1944)

His parents also adopted a child, Josie Humphrey.

Charles's father, James Augustus Fraser (1827-1893), was the second-oldest in a family of 12. His mother, Fanny Maria Rankin (1848-1925), was the eldest of six children.

At the time of Charles's birth, Cookshire was a small but important village in the township of Eaton, one of many townships that made up the "Eastern Townships" of Quebec. Although the area was known to have been frequented by the Abenaki First Nations people much earlier, the first to permanently settle here were United

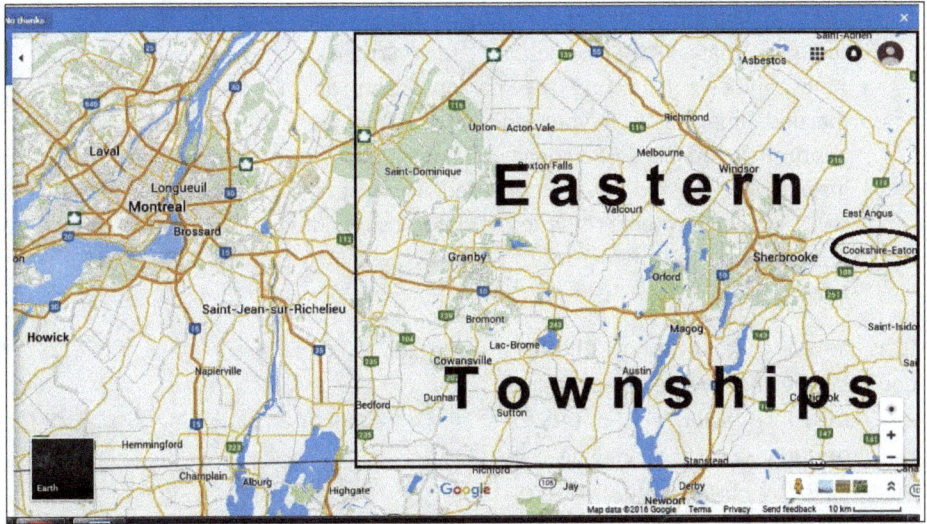

Location of Cookshire in Quebec's Eastern Townships (Google Maps)

Empire Loyalists who emigrated from the northern United States in the late 1700s. Among the settlers were Orsamus Bailey (whose daughter Abigail would later marry Charlie's grandfather, James Fraser) and Captain John Cook, after whom Cookshire was named. During the 19th century, the village gained importance as an agricultural, forestry, military, transportation and political centre, in large part due to longtime local Member of Parliament, John Henry Pope, who served in Prime Minister John A. Macdonald's cabinet. In 1892 Cookshire was incorporated as a town. Charlie would become a life-long resident here, even though his remarkable variety of vocations and interests would frequently take him far beyond. He would become recognized as one of Cookshire's most colourful and unique native sons.

Charles was baptized at Eaton Congregational Church on April 15, 1884. (See his Record of Birth and Baptism in Appendix C.)

The reader may have noted the variation in the spelling of the siblings' family name (Frasier versus Fraser). This dichotomy is illustrated opposite by the signatures of Charlie and his brother James on the same document.

Signatures of brothers James Frasier and Charles Fraser

The origin of this spelling variation is explained by Charlie himself in a 1946 letter to his cousin Mabel Fraser:

> Now, about that "miserable" (wish could find a stronger word) "i" in FRASER: Grandmother Frasier (Abigail Bailey) was from the U.S.A. She thought that FRASER sounded too Scotchy, so she wished to Americanize it. She had the parson insert an "i" when they were married. Imagine it, if you can? All the members of our family were baptized "FRASIER". Personally, I have never used or recognized the inserted "i". Nor did my brother Jared. The rest (retained the spelling), as baptized. Once when writing to me, Jared commented "Grandmother tried to make a bastard name out of the HONORED NAME of FRASER. Was he angry!"

It is noted that some of Charlie's cousins, specifically the descendants of Charles Ira Frasier, also opted for the original Fraser spelling as evidenced by their signatures on historical legal documents.

Signatures of Charles Ira Frasier's three children

Charlie the Boy

Only five of Charlie's siblings survived beyond childhood. They are shown in the photograph below with their mother Fanny and adopted sister Josie. This photo was taken shortly after their father's death in 1893, when Charlie was only 12 years old. On the following page are earlier photos of Charlie's mother and of Charlie with his brother Henry.

```
Likely taken around the time of the
death of James Augustus Frasier
in 1893, as at that time -
(front row) Hattie would have been 7 yrs. old
            Jared      "    "    "  20  "   "
            Nellie     "    "    "  16  "   "
            Fanny
            Henry      "    "    "  13  "   "
(back row)  Josie
            Charles    "    "    "  12  "   "
            James      "    "    "  18  "   "
```

Charlie with his mother Fanny, his siblings, and his adopted sister Josie, circa 1893 (Photo courtesy of Gloria (Frasier) Bellam)

Although little is known about Charlie's boyhood, family folklore says that he was teased in school due to his serious speech impediment – he was unable to get his tongue around words containing the letters "L" and "T," for example. This condition was quite probably genetic in nature since some of his later-generation cousins (including me) were similarly afflicted but had the problem remedied through a simple medical procedure.

Charlie attended school at Cookshire Academy (later known as Cookshire High School and now known as Cookshire Elementary School).

In spite of his disability, he obviously succeeded in school, as indicated by his report cards of General Proficiency in grades Model School II (Grade 8) and Academy II (Grade 10). A pencilled note on his Model School II certificate indicates "1st – 928 marks."

Mother Fanny; brothers Henry and Charles, circa 1884 (Photos courtesy of Gloria (Frasier) Bellam)

Cookshire Academy, circa 1900 (Postcard from the author's collection)

Charlie the Boy

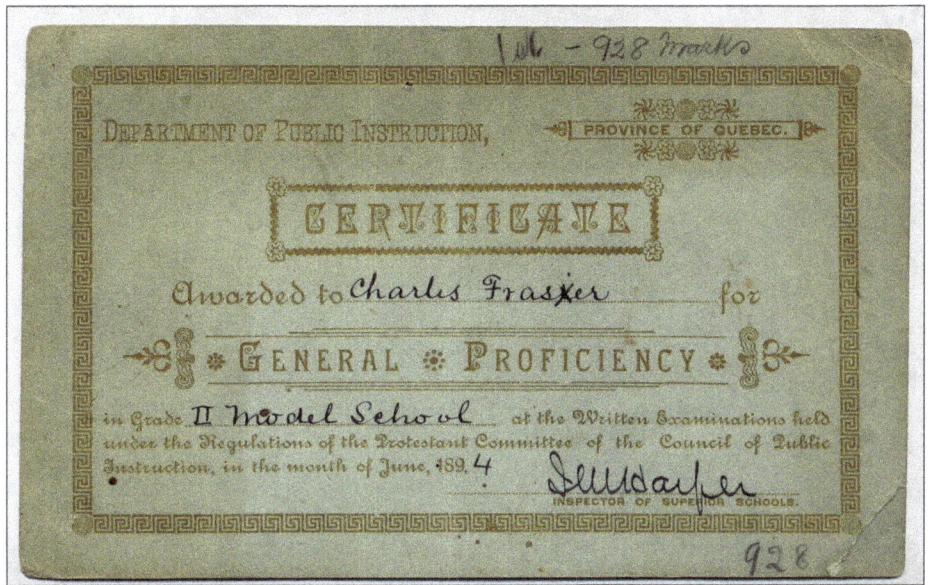

Model School II certificate

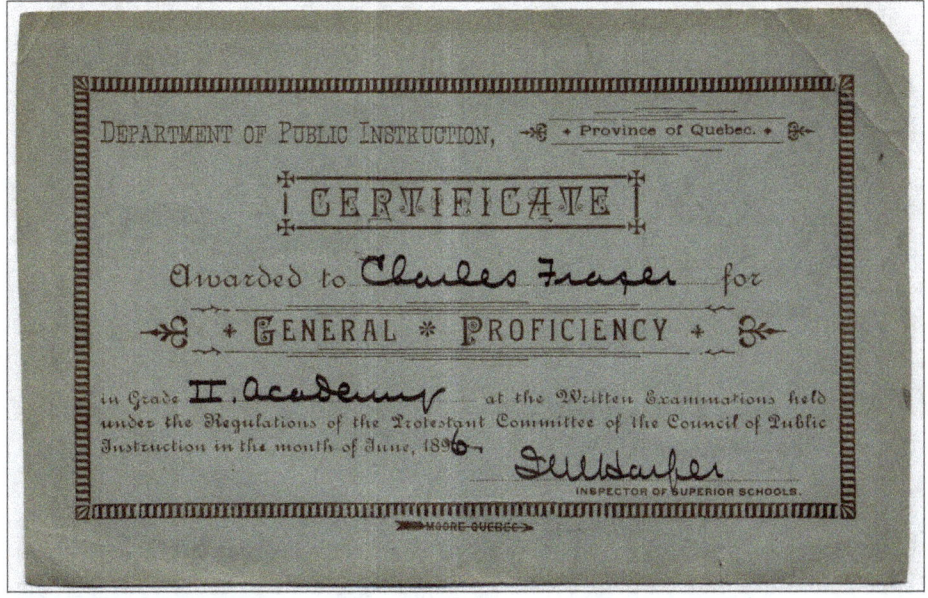

Academy II certificate

In July 1895, "Master Chas. Frasier" was the recipient of a congratulatory postcard from a Maude Ayerst of Dunham, Quebec (presumably his teacher) for succeeding in his school year. It is interesting to note that Maude addressed him as "Dear Charlie," indicating that this diminutive form of his name originated in his boyhood.

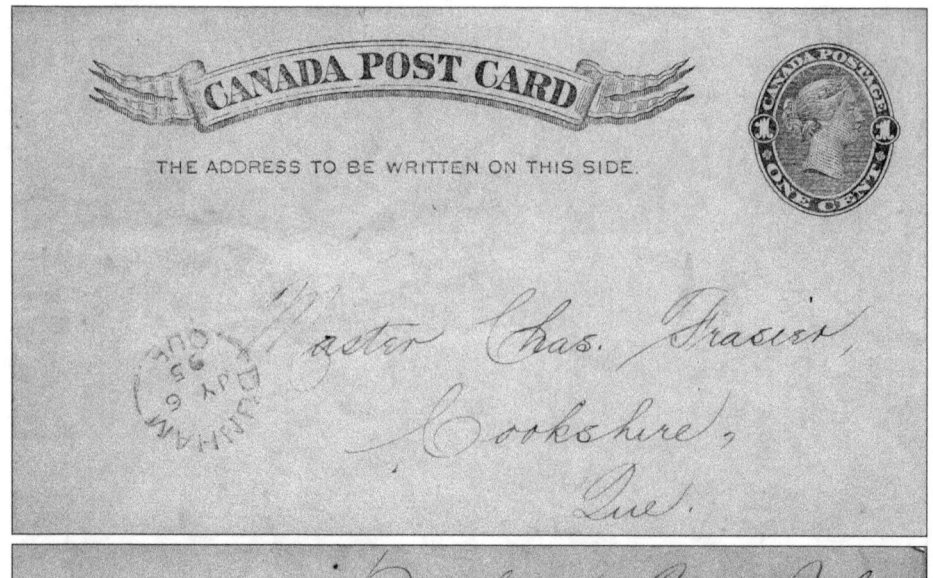

Congratulatory postcard, 1895

Based on his later accomplishments as an adult, it can be reasonably assumed that Charlie the boy was clever, creative and curious. Although his formal education appears to have ended at age 15, he would go on to a life of continuous learning and achievement.

The following chapters of this book, which describe several of Charlie's amazing variety of vocations, are based on a single autobiographical summary sheet provided to me by Charlie (see Appendix A).

Chapter 2   Charlie the Prospector

CANADIAN GOLD HUNTERS
C.C.Fraser,
COOKSHIRE, QUE.

Canadian Gold Hunters letterhead, envelope back

MINING IS THE BACKBONE OF CANADA
GOLD: THE BANK OF CONFIDENCE

Although he never "struck it rich," Charlie was a serious prospector and mining developer. He and his brother Jared ("Jed") Fraser set up a company named Canadian Gold Hunters Reg'd. As manager and secretary-treasurer, he managed their many mining claims in the Eastern Townships, in northern Quebec and in other parts of Canada. They collected ore samples from these holdings and had them assayed for gold and other precious metals content by government laboratories.

The Fraser brothers also prospected for placer gold (i.e., gold found in alluvial deposits – deposits of sand and gravel in modern or ancient stream beds, or occasionally in glacial deposits). It is not known how much gold they found in this form, but it is certain that they did find some. When Charlie dressed up in his Sunday best, the fob of his pocket watch was adorned with a significant gold nugget. As a child, I was very impressed with its

Charlie with pick and shovel outside his mining camp, 1922 (Photo courtesy of Gloria (Frasier) Bellam)

size and colour. It probably weighed about three ounces. At today's prices, it would be worth about $4000 Canadian.

In the early 1920s, Jed and Charlie built a log cabin mining camp in northern Quebec, in the Rouyn area which was just beginning to be developed. Their camp on Fraser Lake (later named Colnet Lake) in Montbray Township was 30 miles from the nearest supply depot. All provisions had to be brought in by packsack, over lakes, through woods and brush and across swamps and muskeg. Charlie and Jed showed hospitality to other gold seekers passing through the area. On many cold winter nights their cabin was filled to its utmost capacity by miners and prospectors, their dog teams tethered in outside sheds. As niece Gloria (Frasier) Bellam recounts: "People were continually calling at the camp for meals and bed while on the trail going to and from their claims, but my uncles were always

Location of Colnet Lake (lac Colnet), Que (Google Maps)

Fraser's mining cabin on Colnet Lake (Photo from the author's collection)

generous-hearted, friendly and hospitable to all, regardless of nationality, colour or creed."

Fraser family folklore contains some very interesting stories of Charlie's time with his brother Jed and cousin Donald (my dad) at that remote cabin on Colnet Lake. One day Charlie developed a toothache, and the pain became unbearable. Since there was no dentist within reach, Charlie picked up a pair of pliers and asked his

Fraser's cabin on Colnet Lake; Charlie's photo caption

cousin to extract the offending molar. When Donald said "I can't do that!", Charlie grabbed the pliers and after much yanking and twisting extracted his own tooth!

On a lighter note, on another occasion, during a late night visit to the outhouse, Charlie was interrupted by a voice calling "Hoo …hoo! Hoo-hoo!" Thinking that it was Donald, he responded "Wait a minute… I'll be there soon." But the voice was very persistent and kept calling "Hoo-hoo!" Charlie was becoming increasingly annoyed with his cousin's impatience. When he finally emerged from the outhouse, he noticed an owl perched on a nearby tree branch!

Artist rendition of Charlie pulling his own tooth (Sketch by James Harvey)

Cousin Charles W. K. Fraser recounts another humorous anecdote. "Once when Charlie was up North visiting Jed, they came upon a bear while walking some six or eight miles deep in the woods. The bear was on the same trail walking directly towards them. Charlie asked Jed: 'What should we do?' Jed calmly replied 'The bear usually veers off the path.' Charlie then asked 'And what if he doesn't?' Jed responded 'Then **we** do!'"

Although an expert woodsman, Charlie nevertheless wisely took precautions to avoid becoming lost while prospecting in the sometimes uncharted geography of northern Quebec. He was constantly aware of his bearings thanks to a pair of compasses – one that he wore on his coat and another that he kept in his pocket.

The area where the Frasers prospected almost a century ago is still the site of gold mining exploration and development today. Globex Mining Enterprises Inc. currently has operations in the Colnet Lake area. The company's website mentions this region from a historical perspective: "Early work on the property likely dates

Jed inside cabin on Colnet Lake

Charlie's compasses (Photo by author)

Top to bottom: Charlie holding a prospector probing bar; mining claim posts on Colnet Lake; Charlie's photo caption

Storage camp on Lake Dasserat; Charlie's photo caption

Artist rendition of owl above outhouse (Sketch by James Harvey)

back to the early 1920's. Very little information is available on this period but it is clear the area was known to have potential, especially for gold."

Later, Charlie's brother Jed built another cabin some 50 miles farther south near the mining town of Ville-Marie (Quebec), very close to the Ontario-Quebec border.

Jed's cabin near Ville-Marie (Photo courtesy of Gloria (Frasier) Bellam)

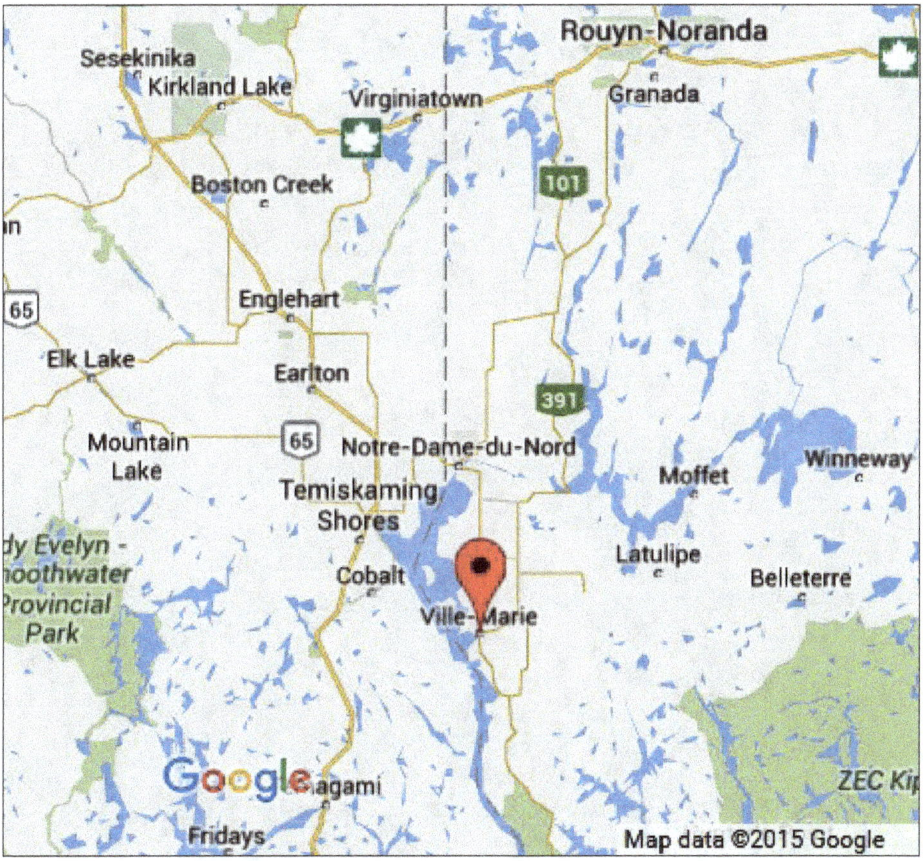

Location of Ville-Marie (Google Maps)

OHIXIHO

Chapter 3   Charlie the Astronomer

Although Charlie didn't describe himself as such in his autobiographical summary, he was nonetheless a bona fide astronomer. Throughout his life, he had a keen interest in natural and physical sciences, particularly in celestial objects and phenomena. Cousin John Fraser recalls how thrilled Charlie the octogenarian was to visit the famous Mount Palomar Observatory during their visit to California in the early 1960s.

Charlie was a Life Member of the Royal Astronomical Society of Canada (RASC). He subscribed to their journal and communicated with the Society, as evidenced by

Mount Palomar (California) Observatory (Photo from Mount Palomar/Caltech website)

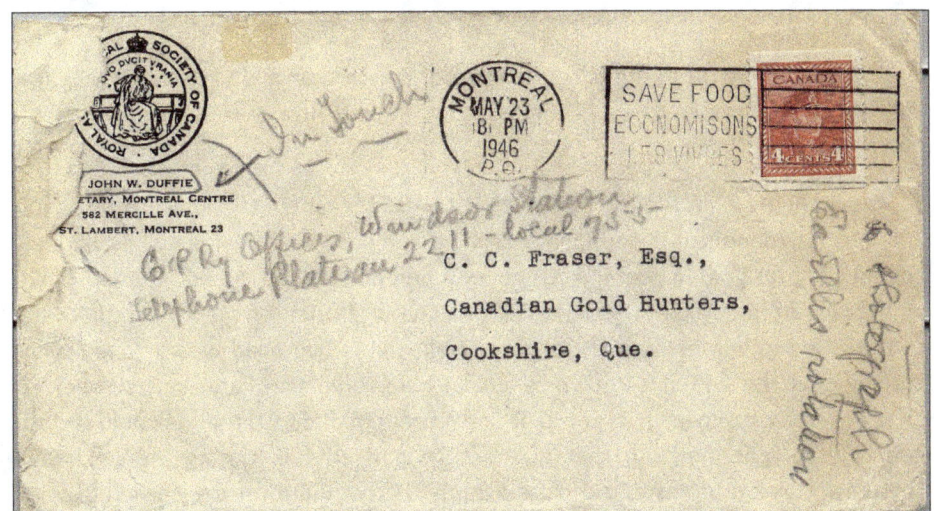

Envelope from Royal Astronomical Society of Canada, 1946 (From author's collection)

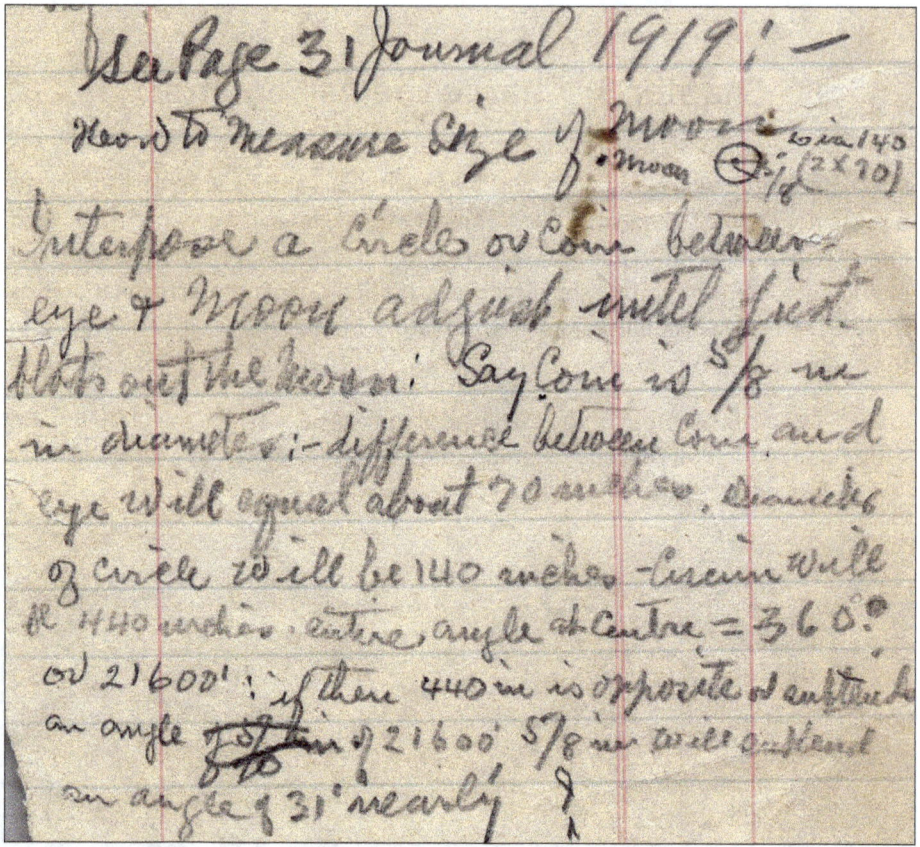

Reference to RASC Journal on how to measure the size of the moon

this 1946 dated envelope. A pencilled note on the envelope reads "to photograph earth's rotation."

Among the handwritten notes contained in the brown envelope that Charlie gave me was this reference to a 1919 RASC Journal about how to measure the size of the moon.

As fascinating as his Mount Palomar visit was, the unquestionable highlight of Charlie's astronomical experiences occurred some 30 years earlier while he was prospecting for gold in northern Quebec. When he stepped outside the log cabin onto the frozen surface of Colnet Lake on the afternoon of March 21, 1930, he looked up and witnessed a most remarkable sight. Overhead was a spectacular display of circles and arcs known as a solar halo complex. Charlie rushed back to the cabin to grab a pencil, a foot ruler and a sheet of paper so that he could record the celestial phenomenon that was occurring overhead. During the next few minutes he wrote notes and drew diagrams that would form the basis of a very detailed document "Fraser's Halo Complex of the Sun (March 21st, 1930)" that he later produced. The principal content of the document is excerpted below. A copy

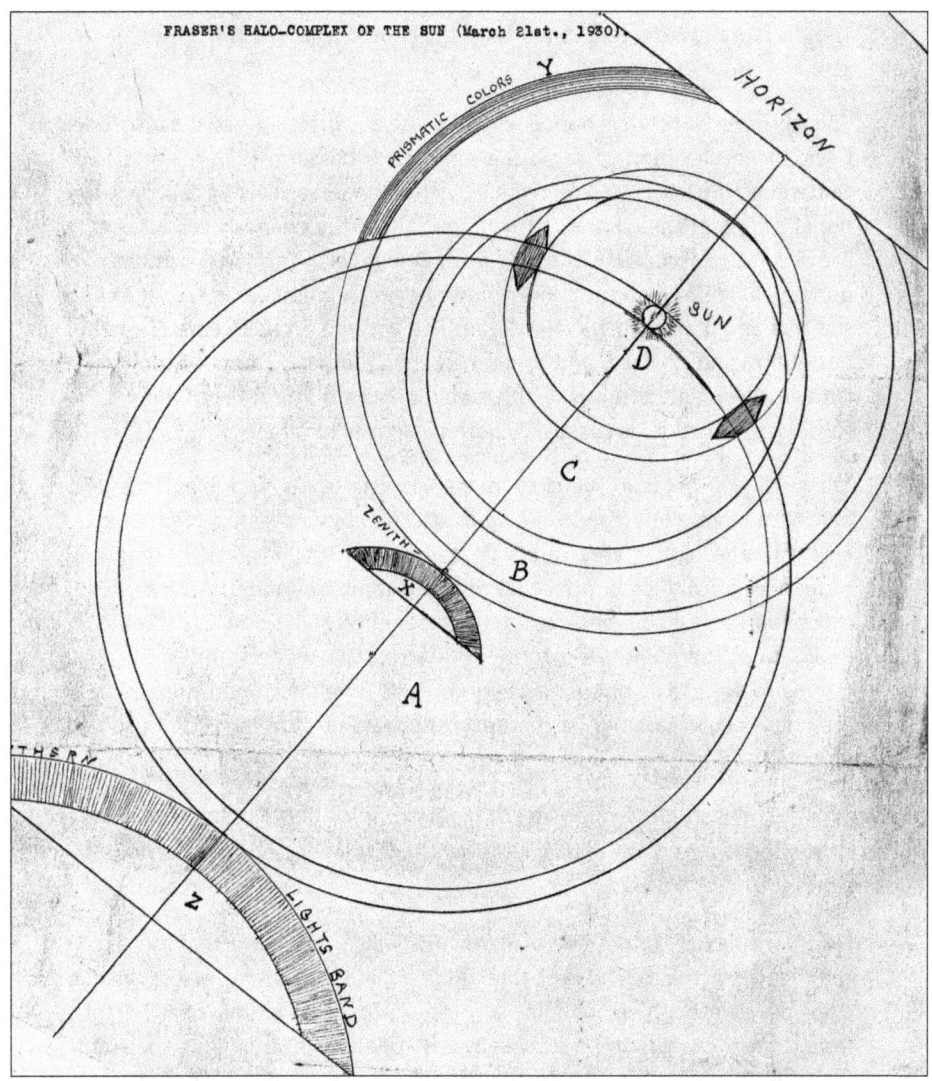

Diagram of Fraser's Halo Complex of the Sun, March 21st, 1930

of Charlie's original unedited documentation is presented in Appendix B.

Excerpts from Charlie's notes (edited):

>A very rare halo complex of the sun was observed by the author, C.C. Fraser, from the snowy frozen surface of Colnet Lake (Fraser's Lake) at the Canadian Gold Hunters' mining camps, Montbray Township, Abitibi County, Province of Quebec, Canada (16 miles N.W. of Rouyn, Que) on March 21, 1930 at 3 hours 50 minutes P.M., Eastern Standard Time. Approximate position at the time of display: Latitude N. 48.12, Longitude W. 79.22. Position of sun S. 60 W. Observed duration, about 40 minutes.

Temperature, 12 degrees F (early in the morning it was 6 degrees below zero F).

The observer's location happened to be at a particularly favourable one, being at an elevation of some 900 feet. Lake Dasserat (913 ft.) and Lake Duparquet (882 ft.) are connected by the Kanasuta River into which the outlet of Colnet Lake (Fraser's Lake) flows. Both the river and Lake Dasserat are about three-quarters of a mile from Colnet Lake. Lake Dasserat has two outlets. One, the Kanasuta River, flows north through a string of lakes and streams and ultimately reaches James Bay and the Arctic Ocean. The other outlet flows southward, and in time reaches the headwaters of the Upper Ottawa River and eventually the Atlantic Ocean. Therefore, Colnet Lake (observer's location) is practically at the height of land.

The author's original sketches made on the spot, and diagrams and pertinent data, duly confirmed and attested by witnesses are the only extant, made of this very rare halo complex of the sun, at the time and place of occurrence. Consider that this halo compares quite favourably with the remarkable halos observed by Gassendi in 1630, and by Hevelius in 1661, and thereupon is worthy of note. We understand that alike atmospheric effects of such magnitude and grandeur and completeness may not occur again during the next 100 years or more, and be observed and depicted.

(Editor's note: The figures opposite are diagrams of the Gassendi and Hevelius halos referenced above, as well as the Loomis halo of 1880 as compared with Fraser's halo.)

N.B. The original sketches were made hurriedly on the surface of the lake in the heart of the bush. Probably there will be scientific errors as the drawings were made within a time limit with crude instruments. Prospectors camps are not usually equipped with the latest scientific facilities. Possibly, some interesting features were omitted or unrecognized. From enquiry, no one in the surrounding area had ever seen anything of the nature comparable in grandeur and completeness.

Characteristics:
- Composed of four circles and an arc, three circles around the sun and one passing through it.
- Apparent diameters of circles A, B, C and D: 48, 40, 28 and 20 inches respectively and proportionally (or 12, 10, 7 and 5)
- Width of rings: ¼ of one inch
- Arc X with 12 inch chord, ¼ of one inch wide band
- Sun's apparent diameter: 1/8 of an inch
- *Above are all perspective measurements taken with a foot-rule at arm's length (18 inches) from the normal eye.*

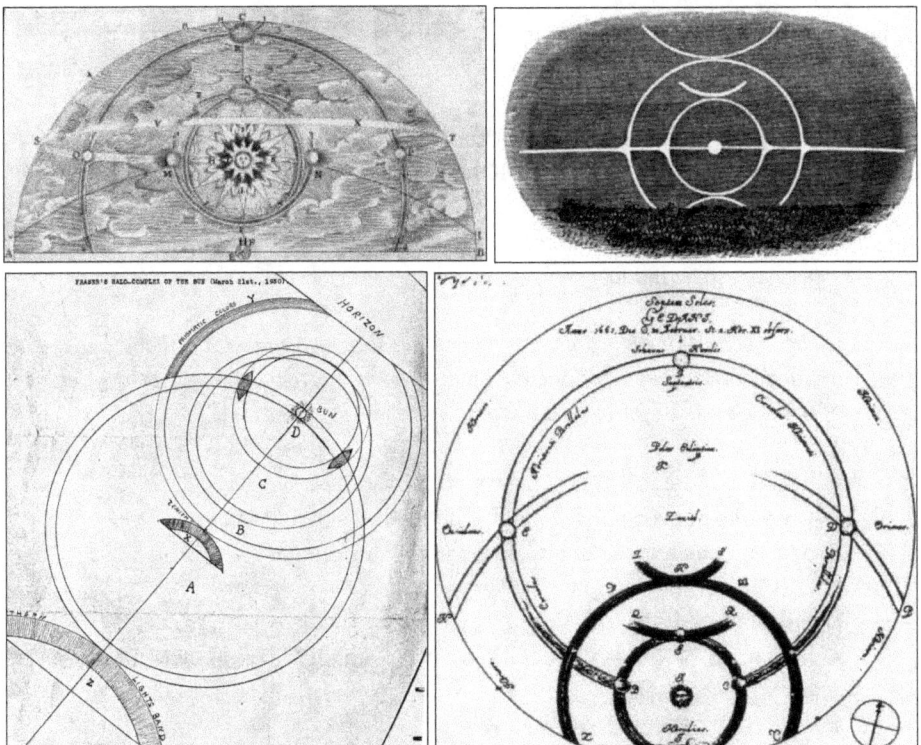

Clockwise from top left: Gassendi halo 1630; Hevelius halo 1661; Loomis halo 1880; Fraser halo 1930. (References: http://www.nature.com/news/2011/110930/full/news.2011.567.html; http://ukazy.astro.cz/halo-historie.php; http://www.photolib.noaa.gov/htmls/wea03012.htm

- Circle A, the largest circle, cut through the sun.
- Circle B, the next smaller circle, appeared to cut circle A through the center.
- Circle C was intersected at the S.W. circumference by the smallest circle D by the width of the ring.
- Circle C is especially interesting, being an irregular halo.
- Circle D had the sun as center.
- Arc X appeared to be in the center of circle A on the circumference of Circle B.
- The ring of the largest circle was of a whitish golden color. The rings of the other circles appeared to be tinged with reddish orange.
- The brightest spots were at the junction points of the largest and smallest circles.
- The arc Y from the south junction of the circles A and B to the horizon was prismatically coloured.
- The arc X was not observed until towards the close of the display, and was of a very reddish colour, and was the last to fade out.

- The center of the largest circle A appeared to be directly overhead of the observer; the other three circles were South 60 W from the observer. All the circles were on the same axis.
- The eastern horizon was practically lake level forest. South and N.W. range of forest covered hills, 30 to 60 feet in height.
- Sun appeared to be in a slight haze at the peak of display; could almost look at the sun directly with the naked eye; sun growing brighter as the display faded.

In the sketches or diagrams, the size of the sun, also the width of the rings, arc and bands have been increased disproportionately relative to the sizes of the circles, in order to display to advantage.

- Position of the sun at occurrence of display: S. 60 W.
- Observed duration: forty minutes (2:50 P.M. – 3:30 P.M.) EST

Appeared to be at peak of display when first observed.

Significance of date:
- March 21, 1930 at 2:50 P.M. Eastern Standard Time (date/time of occurrence)
- March 21, 1930: Vernal Equinox
- March 21, 1930: Last quarter of the moon

Weather Chart:
- March 21, 1930: Early in the morning, 6 degrees below zero. At time of occurrence it was 12 degrees Fahrenheit. The day was clear and bright with N.E. wind of about 20 miles; a few very light scattered clouds; sun set in light clouds; night clear and starry. Preceding few days, light snowfalls and windy. Temperature from freezing to zero.
- March 22: Bright and fine, fleecy clouds; freezing to zero; night clear and bright.
- March 23: Fine and clear all day; sun set in clouds; night starry and clear; temperature zero to freezing.
- March 24: Morning clear and bright; few light clouds; temperature 10 F to freezing; night clear.
- March 25: Similar to the 24th except getting cloudy towards evening, turning to wind and snow; snowed all night.
- March 26: Still snowing, very heavy snowstorm and wind during the night; snowfall of 16 inches.
- March 27, 26, and 29: Fairly mild and cloudy; temperature about freezing.

Coincidently, on the evening of the same day of the halo complex of the sun, there occurred at Colnet Lake at 8:50 P.M. EST, an arched band Z at the N.E. horizon, with a chord of 36 inches; width of band ½ inch, very deeply coloured. Gradually getting wider as it faded, thence sending out streamers over the northern sky. A most magnificent display of aurora borealis. It was first observed at its peak. The center of the arc appeared to be at a point where the N.E. circumference of circle A had been at the time of the phenomenon of the sun and on the same axis.

Extract from the Royal Astronomical Society of Canada Journal 1930 Volume XXIV, page 237:

A SOLAR HALO SEEN AT HAILEYBURY, ONT.

Mr. W. H. Tuke, of Haileybury, Ont., sends the sketch reproduced herewith of a halo seen on March 21 from 4 p.m. until sundown. The mock suns on either side of the sun S were at times too brilliant to look at with unprotected eyes. The outer circle, though incomplete and not so brilliant as the inner one, was very beautifully coloured, especially at the part where it joined the reversed arc. At times it would compare well with a summer rainbow.

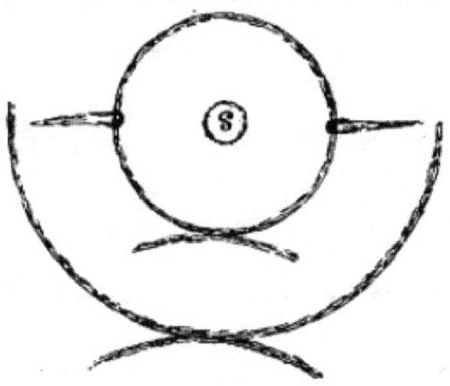

SOLAR HALO
Seen by W. H. Tuke, at Haileybury,
Ont., from 4 p.m. to sundown, March 21, 1930

Halo seen at Haileybury, Ont. 1930 (Reference http:/articles.adsabs.harvard.edu//full/1930JRASC..24..235C/0000237.000.html)

Note: The above described halo was seen the same afternoon as Fraser's halo which was observed on Colnet Lake. Haileybury (Ontario) I would estimate to be some 20 miles S.W.

To protect his rights to the preceding document, Charlie applied for copyright on February 12, 1959. Copyright was granted the following day.

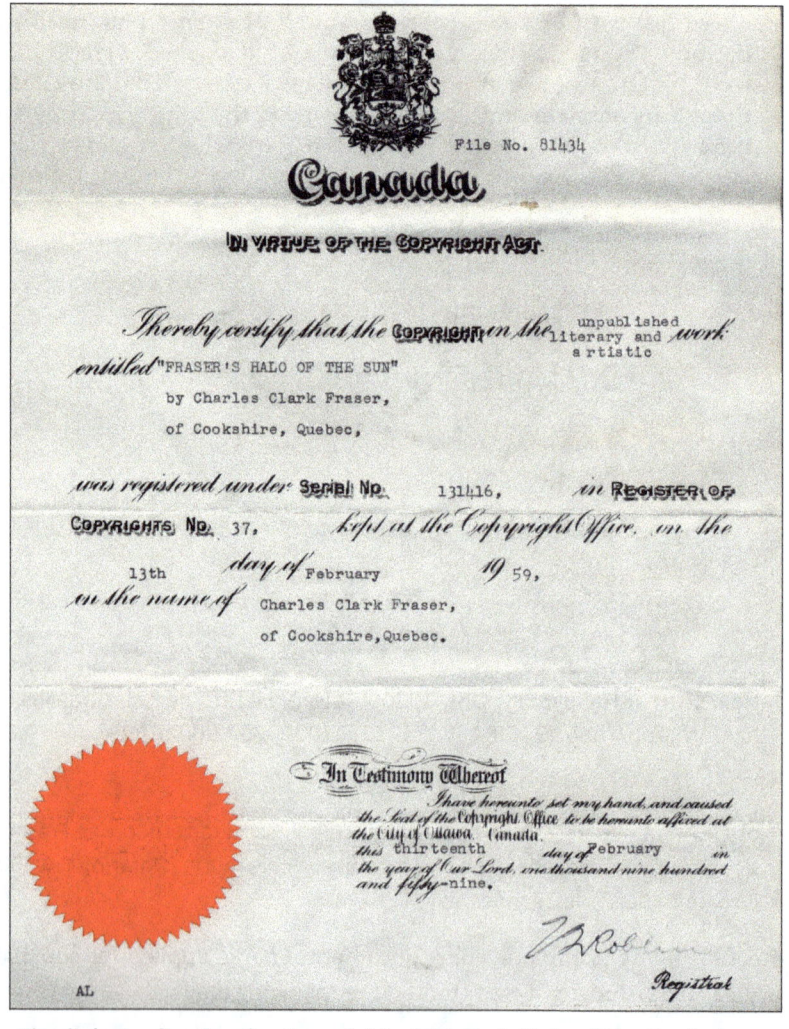

Charlie's Application for Copyright of 'Fraser's Halo of the Sun'; the Copyright Certificate

Chapter 4   Charlie the Inventor

Charles Clark Fraser possessed a brilliant mind that seemed to be in perpetual motion. It was said that he slept only three hours a night. He would spend endless hours, day and night, thinking, designing, experimenting, creating and testing. Nevertheless, one might ask "Was he was really an inventor?"

Thomas Alva Edison, inventor of the incandescent light bulb, defined invention in the following terms: "To invent, you need two things – a good imagination and a pile of junk." Under this definition, there can be no doubt that Charlie was an inventor. Certainly he had a great imagination, as will be demonstrated below. As for Edison's second requirement, anyone who ever visited Charlie's house would attest to the fact that it did contain a pile of junk! Cousin Marilyn (Fraser) Reed, who occasionally did his housecleaning, recalls "Dusting was especially difficult because every piece of furniture was piled almost to the ceiling with so much stuff."

Charlie's "inventions" reflected his broad range of interests. Among the diverse fields of his various experiments and achievements were electricity, radio, television, games, apparel, machinery and literature.

Cousin Malcolm Fraser recalled his dad (Donald Fraser) telling about Charlie's revolutionary electrical experiment in the early 1900s. Charlie succeeded in transmitting electricity over a single ungrounded telephone wire between Donald's farmhouse and his maple sugar camp located approximately a mile away. Although this type of electrical transmission had been the subject of earlier experiments in the United States, it is not clear whether Charlie was aware of them. In any case, his successful experiment might very well have been a first in Canada.

Radio experimentation in this medium's early days was of more than a passing interest to Charlie. In 1922 he constructed the first personal radio in Cookshire. It was a "cat's whisker" radio. According to Wikipedia (https://en.wikipedia.org/wiki/Crystal_radio):

> A crystal radio receiver, also called a crystal set or cat's whisker receiver, is a very simple radio receiver, popular in the early days of radio. It needs no other power source but that received solely from the power of radio waves received by a wire antenna. It gets its name from its most important component, known as a crystal detector, originally made from a piece of crystalline mineral such as galena. This component is now called a diode.

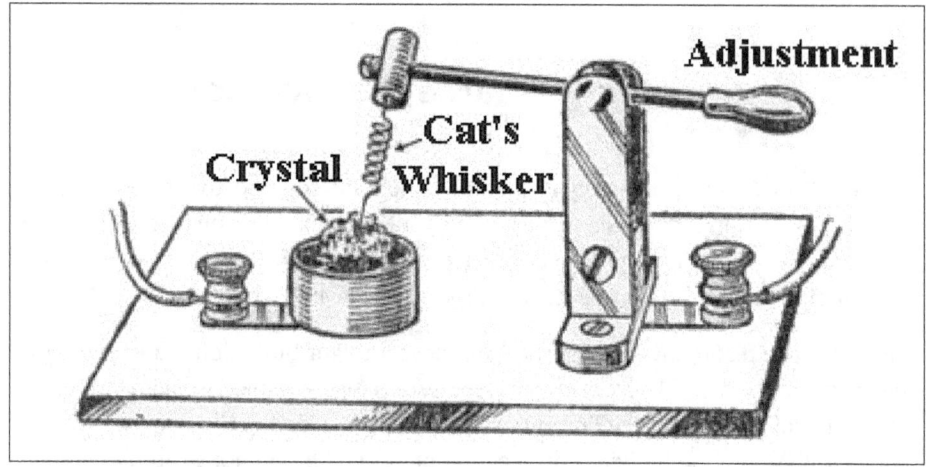

A typical cat's whisker (Illustration from www.historywebsite. co.uk/Museum/Engineering/Electronics/history/earlytxrx.htm)

At that time, there were only a handful of radio stations with regular broadcasting. KDKA in Pittsburgh, Pennsylvania and WBZ in Springfield, Massachusetts were the first commercial radio stations to receive their licences, in 1921 – barely a year before Charlie built his first set. In November 1922 Canada's first radio station, CFCF in Montreal, received its licence.

Although Charlie was over 70 years old when television was introduced in Canada, he enthusiastically embraced the new technology. It is not known exactly when he bought his first TV set, but I recall occasionally going to his house in the late 1950s to watch a special program. Cousin Marilyn (Fraser) Reed recalls how impressive it was to see how, through a system of strategically placed mirrors, Charlie was able to watch TV while eating breakfast in the kitchen even though the TV was in the living room, a full 270 degrees away! Although Charlie noted in his autobiographical summary that he was a "television antenna designer," no details of that activity have been discovered.

Charlie's inventive genius was also applied to the world of games. He devised a checker board with 32 squares instead of the normal 64. An Internet search reveals that J.G. Lallement, a Frenchman, had proposed such a design much earlier, but Charlie was likely unaware of it.

Charlie was not only interested in developing new things, but also took on the challenge of improving an existing invention. Such was the case of the zipper that had come into common use only a few decades earlier. To understand Charlie's involvement, one needs to be aware of the history of this revolutionary invention (from: https://en.wikipedia.org/wiki/Zipper):

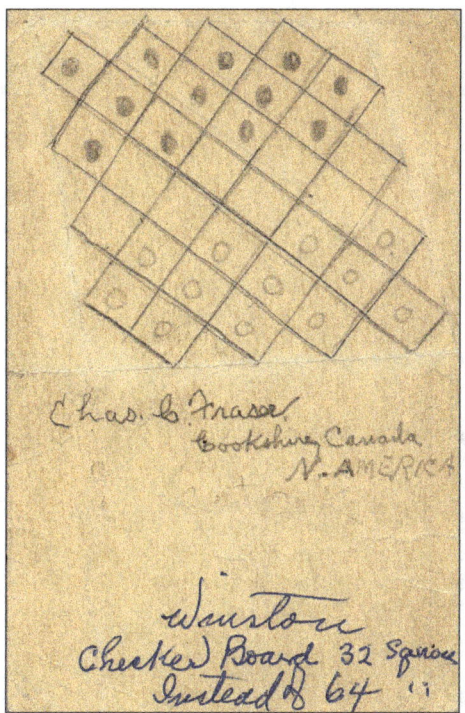

Fraser's 32-square checkerboard

Lallement's checkerboard

The "Fastener Manufacturing and Machine Company" moved to Hoboken, N.J. in 1901. Gideon Sundback, a Swedish-American electrical engineer, was hired to work for the company in 1906. Good technical skills and a marriage to the plant-manager's daughter Elvira Aronson led Sundback to the position of head designer. The company moved to Meadville, PA, where it operated for most of the 20th century under the name 'Talon, Inc.' After his wife's death in 1911, Sundback devoted himself to improving the fastener, and by December 1913 he had designed the modern zipper. The rights to this invention were owned by the Meadville company (operating as the "Hookless Fastener Co."), but Sundback retained non-U.S. rights and used these to set up in subsequent years the Canadian firm 'Lightning Fastener Co.' in St. Catharines, Ont.

In 1945 Charlie contacted Fawcett Publications Inc. of New York to request information about patents relating to the zipper. He received a response written on the letterhead of Talon Inc. of Meadville, Pennsylvaia. However, it is not the letter itself but rather Charlie's handwritten note at the bottom of the letter that is most significant. The note reads: "CCF has new idea." Unfortunately the nature of his new idea remains a mystery. Who knows how the zipper might have evolved had Charlie developed his "new idea"?

> **TALON, INC.**
> MEADVILLE, PENNSYLVANIA
>
> EXECUTIVE OFFICES
>
> August 14, 1945
>
> Canadian Industrial Equipment News
> Gardenvale, Quebec
> Canada
>
> Attention: D. H. Graham, Ass't. Editor
>
> Gentlemen:
>
> Your letter of July 11, addressed to Fawcett Publications, Inc. of New York, N. Y., has been referred to us for reply.
>
> Seemingly, the present form of slide fastener or the zipper as it is more commonly called, was invented by a Swiss woman by the name of Katherina Kuhn-Moos. The basic patents on the zipper have now expired and there have been hundreds of improvement patents granted thereon to various individuals both in Canada and in the United States, some of which have been assigned to various zipper manufacturers in these countries. You can readily understand that we do not know to whom all of these patents have been granted or assigned.
>
> We trust this answers your inquiry.
>
> Very truly yours,
> R. E. Meech
> Assistant Secretary

Letter concerning zipper patents

Some of Charlie's most interesting inventions were in the literary realm. He coined a number of words and copyrighted some of them.

- Weatables
- clock-face
- OHIXIHO (a word that looks the same forwards, backwards, upside down, and in mirror image)
- nemorist ("a person with a poor memory" – accepted by Funk and Wagnall's Dictionary)
- Oliveine (name of a patent medicine)
- Muskalene (name of a patent medicine)
- parsha (unknown)*
- hobo (a homeless vagabond)*

Opposite: Coined the word "Weatables"

Below: Weatables copyright

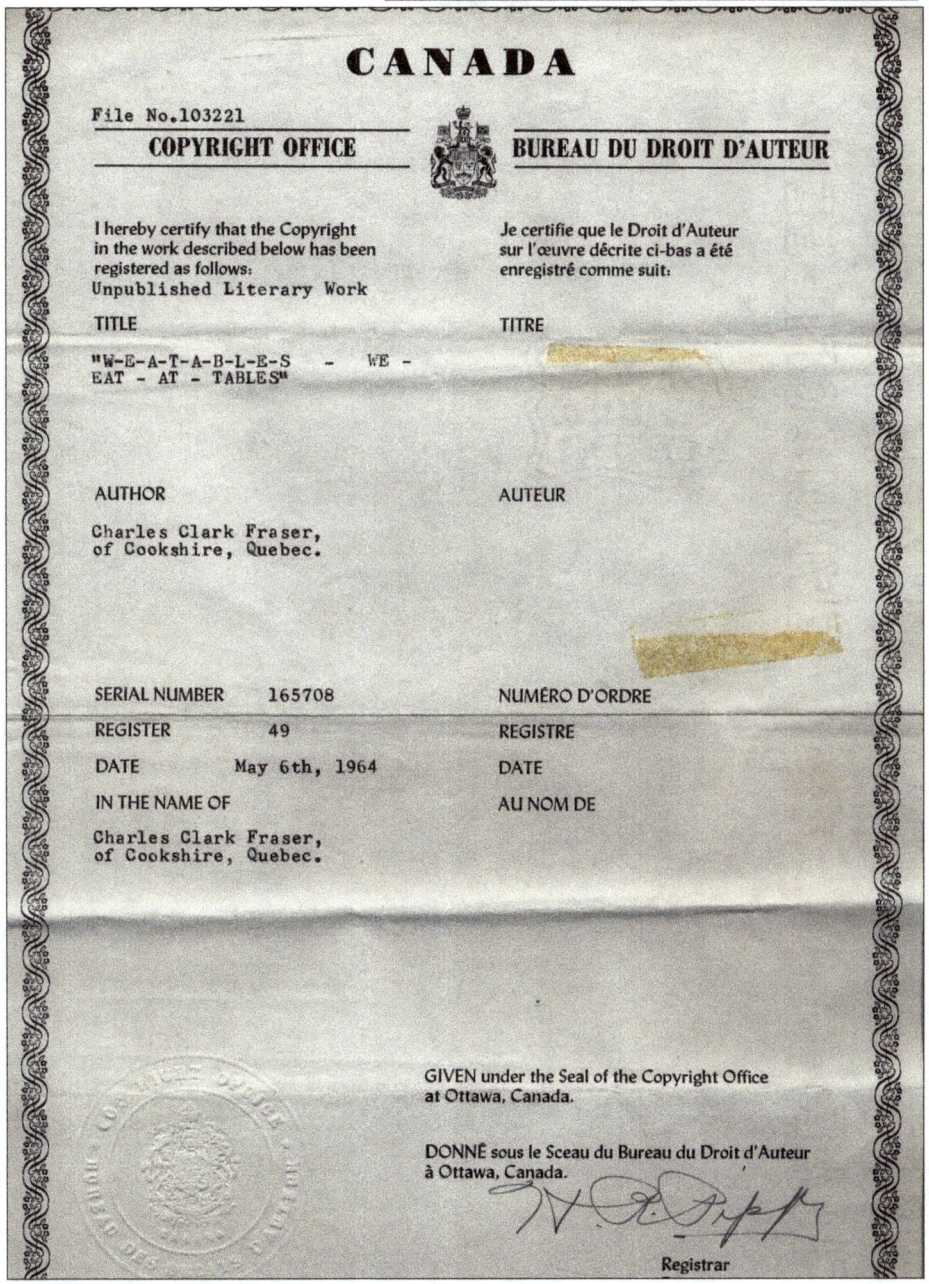

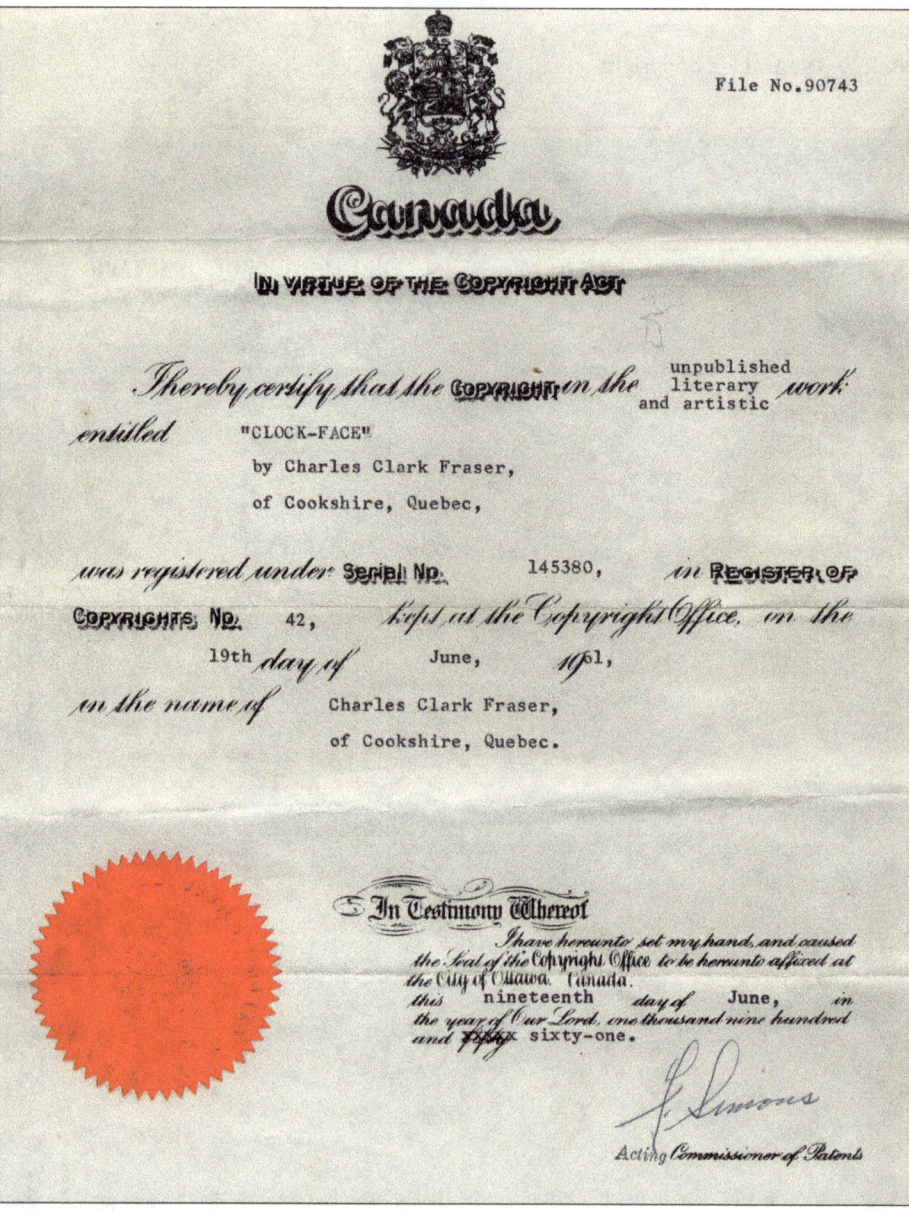

Clock-face copyright certificate

OHIXIHO copyright certificate

*Note: Although listed on his autobiographical summary as words that Charlie had coined, my research questions that claim for these. The word "hobo" came into use in the western USA in the late 1800s, and "parsha" is the Yiddish form for the Hebrew word "parashah" (a portion of the Torah chanted or read each week in the synagogue on the Sabbath). Nevertheless, it quite possible, even probable, that Charlie was unaware of this information. One must realize that there was no Internet and no Google in his era.

Charlie's linguistic interests were not limited to the English language. As a boy, he probably studied Latin at school, as it was a staple of the Quebec school curriculum at that time. In any case, he made use of Latin phrases (e.g. "A POSSE AD ESSE" inscribed on his Canada Centennial monument). It is also believed that he spoke French because many of the clients of the company where he worked (Frasier, Thornton & Co.) were French-speaking. A further indication is that two of the pallbearers at his funeral were French-Canadian friends Pierre Beaulieu and Paul Saint-Laurent. But apparently just being bilingual wasn't sufficient for this man. During the 1950s, while in his seventies, Charlie developed a strong interest in the constructed international language Esperanto, which was gaining in popularity in Europe and the United States at that time. According to the 1963 edition of Encyclopedia Britannica: "A reliable 1952 estimate gave 280,000 users in 32 countries. In the 1960s the total of persons speaking and understanding Esperanto may have approached 500,000."

Speaking this common second (or third) language was seen by its proponents as a means for peoples of different mother tongues to communicate, thereby fostering understanding, peace and harmony – goals shared by Charlie. It is not surprising, then, that just inside his front entrance, Charlie had painted, over the doorway, a large welcome greeting in Esperanto: **BONVENON.**

Charlie's inventive mind and his thirst for knowledge continued well into his old age. He was particularly interested in new advances in science and technology as evidenced by a few of the many 1960s vintage "unusual items" clippings that he would save for my next visit.

In 1963, at the age of 82, Charlie purchased a brand new set of Encyclopedia Britannica to complement his collection of reference books. How he would have appreciated today's Internet reference tools such as Google! Patricia Stevenson Smith, who, as a young person, knew Charlie, concurs: "He was a fountain of knowledge. I can only imagine how he would have enjoyed having a computer with the whole world at his fingertips."

Charlie the Inventor

Ever hear of "lubricated water?" Well, friction between water and solid bodies has been reduced up to 40% by the addition of substances such as guar gum and polyethylene oxide in concentrations of less than 100 parts per million. Benefits, for example, could be the use of smaller pipes for circulating water heating systems, smaller pumps, etc.

Lubricated water

Stream of electronically guided ink droplets replaces typewriter principle in fast print-out system for computers. The typewriter principle is straining to keep pace with the new data processors. The ink droplet technique is like the cathode ray operation. Droplets sprayed from a nozzle are guided to the paper by various electrodes. The ink is attracted to the paper by the very high potential between the nozzles and the platen. The first standard model of this printer is capable of printing 120 characters per second or 1200 words per minute.

Typewriter replacement

Up to 500 pictures on one negative, all the same size as the negative, can be taken by unusual camera. The film is developed in the standard manner but when placed in a special viewer, each of the 500 pictures will appear at the selection of the operator. Resolution is good enough for microcircuit technology.

Unusual camera

New credit card system in Sweden not only pays for the gas, but also starts it pumping. The motorist inserts his card in a slot, dials his card number, then uses the hose to get as much gas as he needs. The motorist's number and amount of gas is recorded on a punched tape which later goes to a computer for billing.

New credit card system

Encyclopedia Britannica set, 1963 edition (Photo by author)

Chapter 5   Charlie the Mathematician

Charlie Fraser was an amazing mathematician. It is said that he could multiply two 6-digit numbers in his head. Not only did he "wow" family and friends with his calculation prowess, but he stumped university professors with some of his math questions.

Among his many mathematical puzzles and formulae were some very practical computing methods. Cousin Charles W.K. Fraser recalls: "He taught me how to easily square any number ending in 5. This is something that I have found useful on many occasions."

Following are some examples of Charlie's mathematical wizardry. In order to allow the reader the opportunity to solve the puzzle below without knowing the solution, the answers are shown upside-down in the subsequent figure.

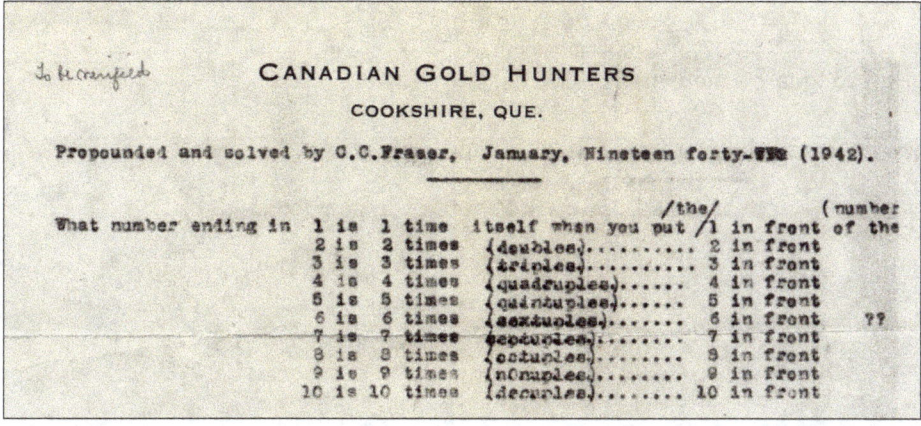

Numerical puzzle

Answers to numerical puzzle above

Note: In the preceding puzzle a pencilled note at the top of the page says "To be verified." In fact, it has been discovered through computer calculations that one of the answers is incorrect. The reader is challenged to determine which answer is in error.

The following three figures describe Charlie's methods for simplifying some difficult calculations.

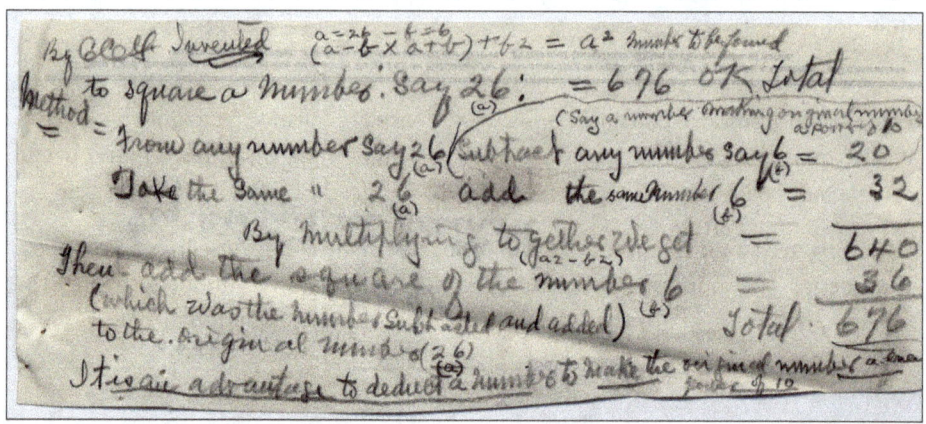

How to square a number

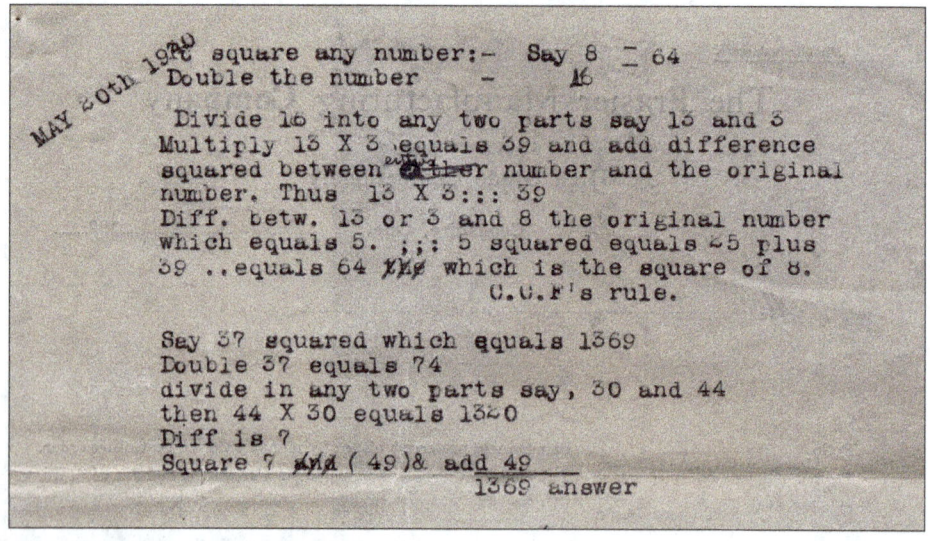

How to square a number

```
To multiply one number by another say, 4 X 7 (28)
Square smaller number, plus the diff. multiplied
by the smaller number. Thus: 4 squared (16)
plus 3 X 4 (12). therefore 16 plus 12 equals 28.
                    or
Square the larger number, minus the diff. multi-
plied by the larger number. Thus: 7 squared (49)
minus  7 X 3 (21) equals 49-21 equals 28.
                  C.C.F's rule.
Say 31 X 45 equals 1395.  1395
    31 X 31 =    961
    14 X 31 =    434     1395            originated
                                            by
Also a² equals b² plus (a+b)(a-b)  Say 87²;;7569
   90 X 84 = 7560 plus 3² = 7569           RULE.
   b equals any number you wish to take  C.C.F.
```

How to multiply one number by another

Charlie seemed to have had a fascination with geometric problems. As indicated by the exchange of correspondence reproduced following, he was able to question and challenge mathematics professors at some of North America's top universities.

```
                     C.C. FRASER,


                                        March 1st. 1944.


To the Professor of Mathematics,
McGill University, Montreal, Que.

Dear Sir:
         Will you kindly forward me the formula for bisecting
(meaning, to divide into two equal area parts) a given circle
by part of the circumference of another circle of equal or un-
equal diameter.
         Also formula for bisecting any given parallelogram or
triangle by circles of different diameters.
         Will very much appreciate this information. Thanking
you, I am,

                                   Yours very truly,

                                   C.C. Fraser.
```

Letter to McGill (and other universities) re: geometry problem

Faculty of Arts and Science,
Department of Mathematics.

McGILL UNIVERSITY
MONTREAL

March 3rd, 1944.

Mr. C. C. Fraser,
c/o Canadian Gold Hunters,
Cookshire,
P. Que.

Dear Mr. Fraser:

    Professor Sullivan has asked me to reply to your letter of March 1st concerning the bisection of certain areas by means of circular arcs.

    It is disappointing for me to have to tell you that there is no algebraic formula (and it can be proved that none can exist) and no ruler-and-compass construction for performing the operation required by the first problem you mention. The statement of the second problem is not quite clear, but from what I judge you to mean, the same observation would apply to it.

    The first problem can certainly be solved by approximate methods to any desired degree of accuracy, but it is useless to try to find an explicit algebraic formula, because the task would be bound to fail.

Yours very truly,

W. Bruce Ross

(W. Bruce Ross)
Assistant Professor
of Mathematics.

Response from McGill University re: geometry problem

Queen's University, Kingston, Ont., March 25, 1944

Mr. C. C. Fraser,
Cookshire, Que.

Dear Sir,

Your inquiry, which was misdirected to Queen's University, Toronto, has been readdressed here.

With regard to the bisection problems which you mention, there is no simple formula in any of them which will give the data for the bisecting circle. The problem of the parallelogram or triangle has too many variables to allow of any one answer. It depends on the shape and size of the figure, the centre of the circle, and the radius of the circle. If the shape and size of the parallelogram or triangle and the position of the centre of the circle were given, it would be a determinate problem to find the radius but the problem would not be a simple one unless the centre were in some convenient position. No one formula could cover all cases.

In bisecting the area of a circle by another circle, if the radii were given it would be a determinate problem to find the distance between their centres. That this is not easy and leads to no simple formula becomes evident if you consider the simplest case -- that of two equal circles. Here the required distance between the centres of the two circles is

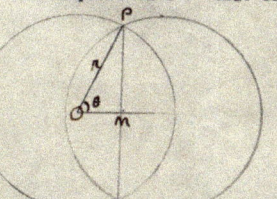

$$2 \, OM = 2 \, r \cos \theta$$

where θ is determined from the equation

$$r^2 \theta - r^2 \sin\theta \cos\theta = \tfrac{1}{4}\pi r^2$$

or $\quad 2\theta - \sin 2\theta - \tfrac{\pi}{2} = 0.$

This equation can be solved by approximate methods and gives $\theta = 1.155$ radians approx., whence $2 \, OM = 2 \, r \times 0.4039 = 0.8078 \, r.$

Yours very truly,

N. Miller

Response from Queens University re: geometry problem

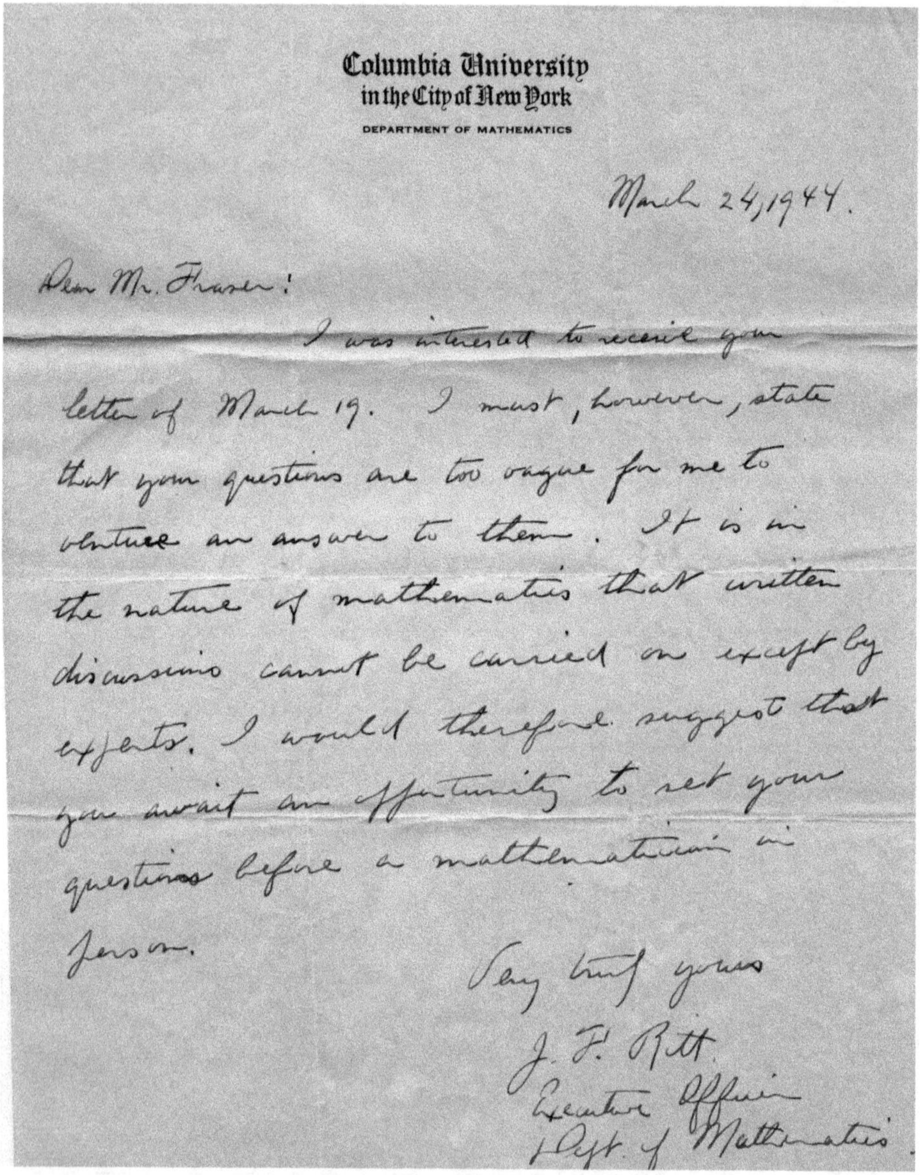

Columbia University
in the City of New York
DEPARTMENT OF MATHEMATICS

March 24, 1944.

Dear Mr. Fraser:

I was interested to receive your letter of March 19. I must, however, state that your questions are too vague for me to venture an answer to them. It is in the nature of mathematics that written discussions cannot be carried on except by experts. I would therefore suggest that you await an opportunity to set your questions before a mathematician in person.

Very truly yours,

J. F. Ritt.
Executive Officer
Dept. of Mathematics.

Response from Columbia University re: geometry problem

Charlie's great fascination with mathematical problems was no doubt further stoked by some mathematical puzzles booklets found among his documents.

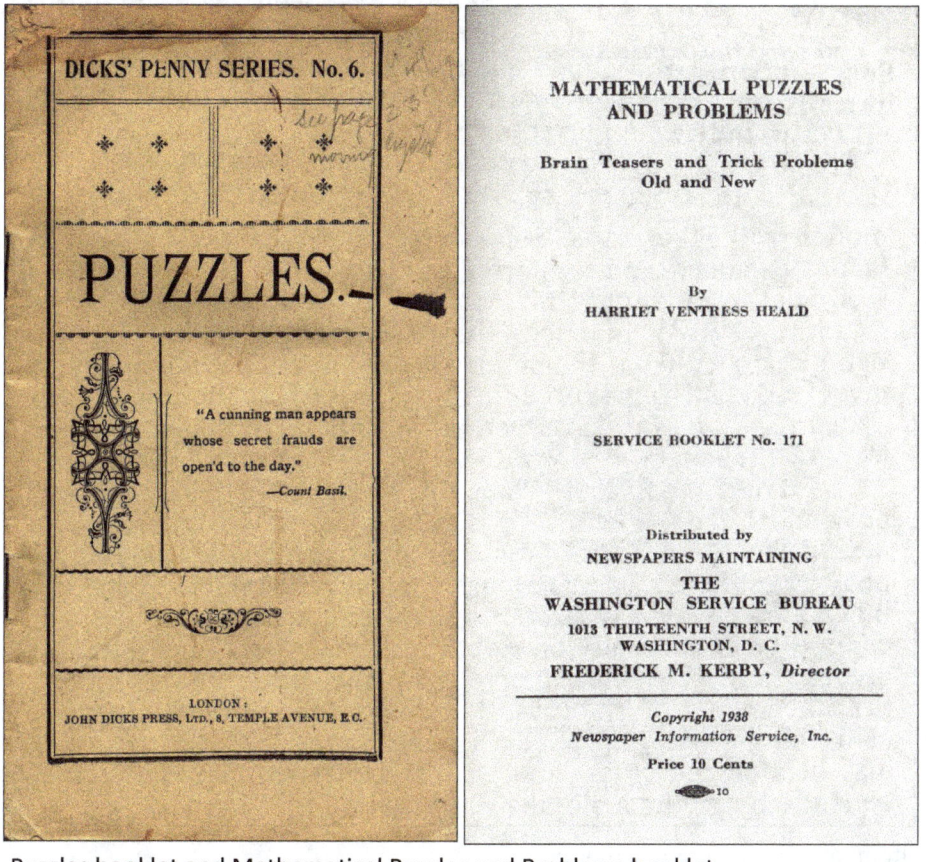

Puzzles booklet and Mathematical Puzzles and Problems booklet

In the case of the Puzzles booklet, Charlie makes reference on the cover page to an item on page 23 relating to perpetual motion.

In the Mathematical Puzzles and Problems booklet, one of the mathematical problems (see figure next page) obviously caught Charlie's attention. His pencilled margin notation indicates that he had a better solution to the problem, but in true Charlie Fraser fashion, he leaves you wondering what that better solution was! The reader is challenged to solve this little mystery.

47.—NEW PERPETUAL ROTARY MOTION.

By an accidental occurrence, it has been discovered that a piece of rock crystal, or quartz, cut in a peculiar form, produces, upon an inclined plane, and without any apparent impetus, an extraordinary rotary motion, which may be kept up for an indefinite period of time.

The crystal has six sides, and being cut accurately from the faces to a perfect convex surface, and held parallel, no motion will take place, because the centre of gravity of each face is balanced and supported in this position of the plane surface; but if a slight inclination be given to the plane, a rotary motion commences, in consequence of the support being removed from the centre of gravity. The impetus once given, the centrifugal force increases the rotary motion to such a degree as for an observer to be unable to distinguish the form of the crystal.

To Produce the Effect.—Place the crystal on a plate or piece of window glass, a china or glazed plate, or any smooth surface, perfectly clean, as grease or a particle of dust would impede its motion. Wet the surface, and give the plane a slight inclination; when, if properly managed, a rotary motion will commence, which may be kept up for any length of time by giving alternate inclinations to the plane surface, according to the movements of the crystal; to heighten the pleasing effect of which a variety of paper figures, harlequins, waltzers, &c., may be attached. The first trial of the experiment had better be made by giving a slight rotary motion to the crystal.

Item relating to perpetual motion

---

10. Given a three-quart measure and a five-quart measure, how would you draw seven quarts of water?

Solution. Fill the three-quart measure and pour it into the five-quart measure. Fill it again and pour the water into the five-quart measure until the five-quart measure is full. There is now one quart left in the three-quart measure. Empty the five-quart measure and pour the one quart from the three-quart measure into it. Refill the three-quart measure and pour it into the five-quart measure. Fill the three quart measure again and you will have seven quarts, four in the five-quart measure and three in the other measure.

Mathematical problem for which Charlie had a better solution

Chapter 6 Charlie the Athlete

Charlie never competed in the Olympic Games, the Boston Marathon or the Tour de France but he certainly could have. He was an outstanding athlete. He possessed a trifecta of athletic traits – strength, speed and stamina – that he used to their absolute limits. If not for the meticulous records that he kept, little would be known of his athletic exploits.

For each distance that he ran (or walked or biked, as the case may be) Charlie recorded the distance and the date together with the time it took, down to the fifth of a second. All runs were self-timed using his Swiss-made Park stopwatch, now in my possession. To validate the accuracy of the stopwatch, a 30-minute test was recently performed to measure the stopwatch timing against computer timing. The computer time recorded was 30 minutes and 1.15 seconds resulting in stopwatch accuracy of greater than 99.9% (or less than one one-hundredth of a second per 10 seconds). This result provides a very high level of confidence in the accuracy of the times recorded by Charlie.

Charlie's Swiss-made Park stopwatch

Charlie's running/walking notes for 1910 (when he was 29 years old) are particularly interesting.

Among the log entries for 1910 are:

Apr. 16	quarter mile run	1 minute 12–1/5 seconds
June 02	100 yard dash	11–1/5 seconds
June 15	2 mile walk	18 minutes
June 27	14 miles on (railroad) ties	4 hours 20 minutes

But most striking of all is the entry for September 26. It reads as follows:

| Sept. 26 | 100 yard dash | 9–1/5 and 10–1/5 |

A notation beside the entry reads: "at the time world record was 9 1/5 seconds for

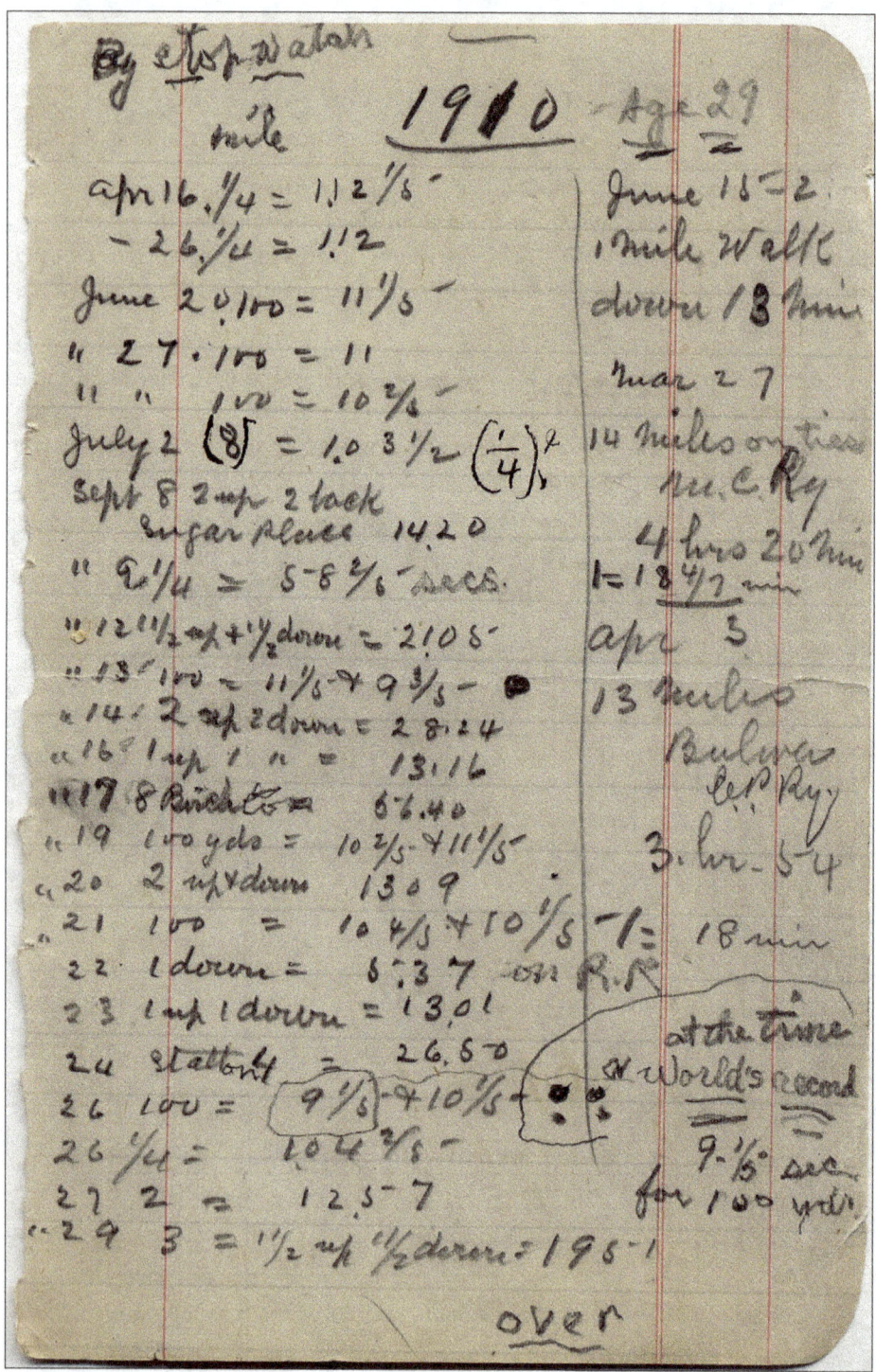

Log of running times, 1910

100 yards." In fact, Internet research shows that the **official** world record in 1910 was 9.6 seconds. (Faster times had been recorded but were not declared as official world records.) It was not until more than 50 years later that the official world record dropped to Charlie's time of 9.2 seconds. The great Canadian sprinter, Harry Jerome, set that record in 1962.

Athlete	Year	Time (sec)	Remarks
F. C. Saportas	1870	10.5	Official world record
Horace H. Lee	1877	10	Equalled official world record
W. C. Wilmer	1878	10	Equalled official world record
Lon Myers	1880	10	Equalled official world record
Arthur Wharton	1886	10	Equalled official world record
J. Owen, Jr.	1890	9.8	Official world record
D. J. Kelly	1906	9.6	Official world record
Eddie Tolan	1929	9.5	Official world record
Frank Wykoff	1930	9.4	Official world record, without starting blocks
Jesse Owens	1933	9.4	Equalled world record, set US high school record
Mel Patton	1948	9.3	Official world record
Ken Irvine	1961	9.3	Equalled professional 100-yard world record
Harry Jerome	1962	9.2	Record made Jerome the only athlete to own both the 100-yard and 100 meter world record simultaneously.
Frank Budd	1962	9.2	Official world record

Official world record holders for 100 yard dash (Reference: https://en.wikipedia.org/wiki/100-yard_dash)

Other examples of Charlie's running, walking and biking notes (for 1927 to 1933) are shown in figures following. It is noted that on August 15, 1933, at age 52, he ran 100 yards in the remarkable time of 10-3/5 seconds!

Charlie's September 17, 1936 note sheets indicate that even at age 55, he could still burn up the dirt track at the Cookshire fairgrounds, running the 100 yard dash in 12-1/5 seconds. But this page contains some interesting additional data. Because he was running on a muddy track he was able to count the number of strides (i.e., steps) and hence the length of his stride. This analysis led him to question which was most important, stride length or speed (stride frequency). Recent analysis of world champion sprinter Usain Bolt's running characteristics (reference Journal of Human Kinetics, March 2013) indicated that his increased stride length differentiated him from his closest competitors. Since Charlie's stride length (5.12 feet) was considerably shorter than Bolt's (8.27 feet), the obvious way to better his time would have been to increase his stride length.

Running, walking and biking notes, 1927-1930

Charlie the Athlete

Running, walking and biking notes, 1933

> Sept 17 – 1936 Exhibition G.
> Age 55 See
>
> Run 100 yds — 12 1/5
> (foot-(prints)?
> 1st. 50 yards in 28 1/2 strides
> 2nd. 50 yards in 30 strides
> Again: –
> (foot-(prints)?
> 1st. 50 yards 28 strides
> 2nd. 50 yards 29 1/2 strides
>
> 1936
>
> mud showed foot prints
>
> See length of stride +
> over 10 feet
>
> Comparison with runners,
> about 10 feet to stride
> See how to increase stride
> or speed of stride

Running notes, September 17, 1936

```
FOOT-PRINTS ON THE SANDS OF TIME
By a Man of 55 Years, Sept.17,1936

Distance                        Time

100 yards      ...1st.Trial     12-1/5    Seconds
1st.50 yds.    ............     28-1/2    STRIDES
2nd.50 yds.    ............     30        STRIDES


100 Yards      2nd.Trial        12-1/5    Seconds
1st. 50 yds.   ............     28        STRIDES
2nd. 50 yds.   ............     29-1/2    STRIDES


     IMPRINTS, Shown on a MOIST DIRT TRACK. Straight


PROBLEM:                      of      STRIDE
         OUGHT THE LENGTH or THE XXXXX OR
         THE XXXXXX BE INCREASED ?
              On the   SPEED
         HEIGHT: 5'- 10"
         WEIGHT  136
         LEG, inside: 30-1/2 ins.

                    C. C. Fraser
```

Running notes, September 17, 1936

Charlie continued running well into his eighties. In 1967, at age 86, he applied for his amateur card from the Amateur Athletic Union of Canada. In a letter dated August 9, 1967, his request was granted.

Perhaps one of the reasons for Charlie's athletic abilities was that he had an "athlete's heart." According to the United Healthcare website (www.uhc.com):

> . . . athlete's heart itself is benign. It's simply the term used to describe a heart that is enlarged, with thickened muscle walls – the results of intense athletic training. Those physiological changes allow the heart to pump more blood per heartbeat, and often results in a slower, stronger pulse, notes the Merck Manual, the widely respected medical textbook.

No. 10004 Q
Amateur Athletic Union of Canada
Quebec Branch

This is to certify that Mr. Charles C. Fraser of Cookshire P.Q. is registered as an honorary member with Amateur Athletic Union of Canada as an amateur.

This card expires September, 19__ **AD INFINITUM**

Matthew Dresser, Registrar

Amateur Athletic Union of Canada membership card content

AMATEUR ATHLETIC UNION OF CANADA
UNION ATHLÉTIQUE AMATEUR DU CANADA

FOUNDED
FONDÉE EN 1889

No. 10004 Q

Quebec Branch   Succursale du Québec

August 9, 1967.

Mr. Charles C. Fraser,
P.O. Box 46,
Cookshire, P.Q.

Dear Mr. Fraser-
    Thank you very much for your letter and request for an Amateur Card of the Amateur Athletic Union of Canada.
    We are indeed very honored to welcome you into our ranks, and hope that you accept the enclosed membership card with the keen sense of sportsmanship that you are exemplifying.
    It is indeed gratifying to note that a man of 86 yrs. of age, is still interested in the practice of sports, and I sincerely hope that at some future date you may be able to attend one of our executive meetings, and be able to impart your enthusiasm to some of our younger members.
    Wishing you many more years of health and good luck,
I remain,

                                                Sincerely yours,

                                               Matthew Dresher,

                                               Registration Chairman,
M. Dresher,   Quebec Branch A.A.U. of Canada.
6836 Somerled Ave.,
Montreal 29.
484-6310.

Letter from Amateur Athletic Union of Canada, 1967

Charlie the Athlete

It is believed that Charlie's resting heart rate was very low -- between 30 and 40 beats per minute as compared to a normal pulse rate of between 60 and 100. The Globe and Mail reported on August 13, 2001, that "The slowest resting rate ever recorded by medical science was 28, recorded by a runner." However, London's Daily Express reported a new world record on May 15, 2014: "Exercise enthusiast Daniel Green, 81, could not believe it when doctors told him his heart rate had dropped to just 26 beats per minute."

Given his obvious running talent, it is strange why Charlie didn't race competitively. As far as is known definitively, he only raced against himself and against the clock. Although family folklore says that Charlie once entered a road race against the great Gerard Côté (four-time Boston Marathon winner) and beat him, no evidence has been found to corroborate such a Cinderella story. Nevertheless, the humble Charlie probably dreamed of winning a competition, as expressed in the poem "Run Well!" that was found amongst his clippings.

Stamina was another hallmark of Charlie the athlete. He once walked 65 miles without stopping. His note regarding this extraordinary feat of endurance says simply "Walked 65 miles at one stretch." No date or context is mentioned. However, it might relate to cousin Charles W.K. Fraser's recollection of one of Charlie's long-distance hikes. "One day, he walked all the way to Lake Megantic (a distance of about 50 miles). I guess Sherbrooke (15 miles) wasn't enough of a challenge. He knew he could do it, so he just went ahead and did it."

RUN WELL!

Frank Horne, a young Negro athletic coach in a Georgia school, addressed these lines to a member of his track team:

> Live
> As I have taught you
> To run, boy.
> It's a short dash;
> Dig your starting holes
> Deep and firm;
> Lurch out of them
> Into the straightway
> With all the power
> That is in you;
> Look straight ahead
> To the finish line;
> Think only of the goal.
> Run straight;
> Run high;
> Run hard;
> Save nothing—
> And finish
> With an ecstatic burst
> That carries you
> Hurtling
> Through the tape
> To victory.
> —From "Singers in the Dawn."

Poem "Run Well!"

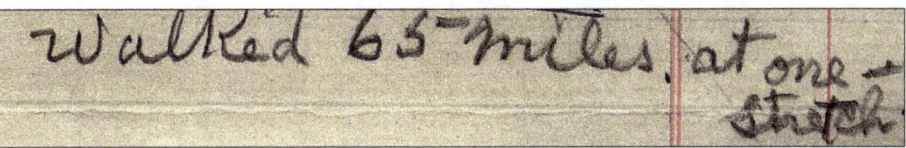

Note re: walking 65 miles

As a young lad, I was party to and participant in another illustration of Charlie's exceptional stamina. Using a two-man crosscut saw, he sawed completely through a three-foot-diameter maple log without stopping once for a rest. I was assigned to handle the other end of the saw, and was practically dead by the end of the ordeal!

Charlie was an exceptional athlete not only because of his speed and his stamina. He was also a man of remarkable strength. Cousin Malcolm Fraser remembers his dad telling about the time that Charlie played "horse," hauling a loaded hay wagon around the fields. But this was child's play in comparison to another of his demonstrations of strength. One of Charlie's handwritten notes contains the following almost unbelievable statement: "C. C. Fraser lifted from the floor 651 lbs with one hand six hundred & fifty-one pounds." This amazing achievement compares very respectably with the exploits of famous strongmen not only of his era but even of today. It is all the more remarkable given Charlie's slight stature (5 ft. 10 in. and 140 lb.) as compared to the strongmen mentioned below, who all weighed more than 250 lb.

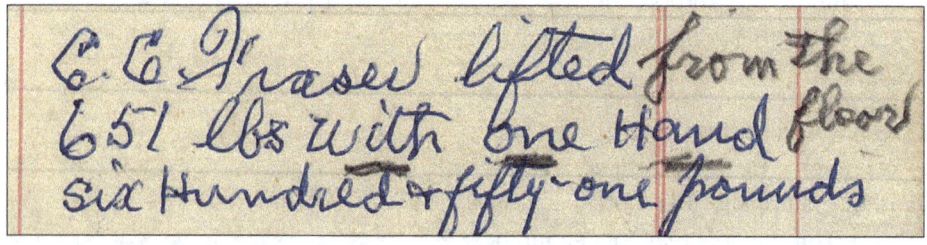

Note re: one-handed lift

Around 1896, the great Quebec strongman, Louis Cyr, did a one-handed deadlift with a dumbbell weighing 525 lb. (reference: https://en.wikipedia.org/wiki/Louis_Cyr). In 1920, German strongman Herman Goerner performed an unofficial one-hand deadlift of 727½ lbs. The one-handed lift record has been regularly broken in more recent times. In 1983, a one-hand lift contest was held in Britain to find a man from the building trade who could lift the greatest number of building blocks with one hand. Clive Lloyd took the day, with a lift of 669 lbs. The current record is 705 lbs. lifted by Swedish strongman Benny Wennberg with his left hand at the European Grip Championships in August 2004 in Loddekopinge, Sweden (reference: http://www.davidhorne-gripmaster.com/article5.html).

Although there is no evidence Charlie was in any way a fighter or a brawler, his incredible strength would have been a huge deterrent for anyone wanting to pick a fight. In addition, he had some interesting tricks in his arsenal. For example, according to cousin Charles W. K. Fraser, he was able to tie a man to a tree without any rope, chain or other type of tie. He apparently learned the technique

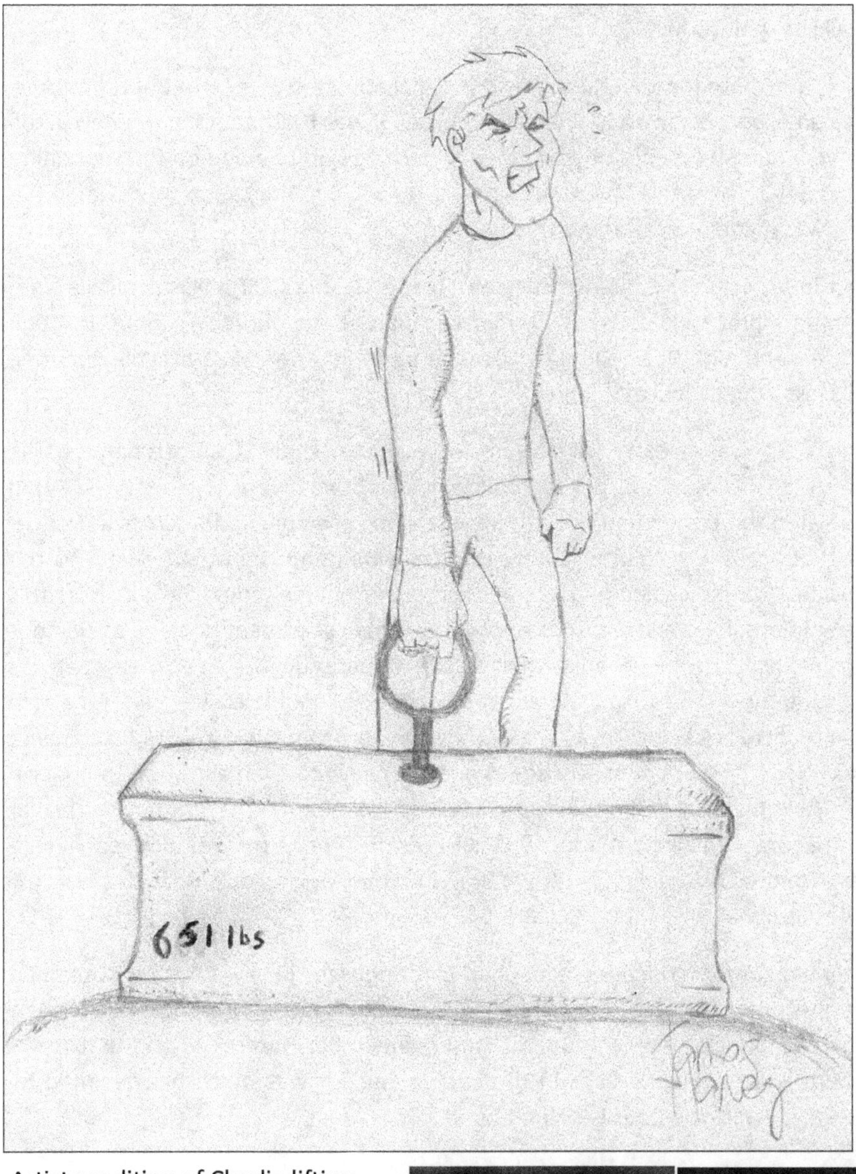

Charlie the Athlete

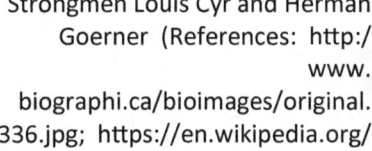

Artist rendition of Charlie lifting 651 lb. (Sketch by James Harvey)

Strongmen Louis Cyr and Herman Goerner (References: http://www.biographi.ca/bioimages/original.336.jpg; https://en.wikipedia.org/

from a First Nations man.

Charlie participated in several team sports including lacrosse, cricket and hockey. He was an expert batsman in cricket. In hockey, I was told that Charlie often acted in the very dangerous role of goal judge, where he had to stand on the ice behind the goal net in the direct line of fire! This is yet another indication that Charlie the athlete was made of very tough stuff.

Niece Gloria (Frasier) Bellam recalls her Uncle Charlie's athleticism on ice with more than a hint of jealousy: "I remember him at the old local (Cookshire) rink skating around with Rosie Staples, also a good athlete, on his arm while ignoring me, his less athletic niece!"

Although he was an excellent athlete in so many respects, surprisingly Charlie could not swim. Apparently it was because his bones were too heavy. My dad once related the story of being with Charlie when they needed to cross a body of water that was more than 6 feet deep. While Dad and the others swam to the other side, Charlie walked across underwater! Although some might dismiss this as being impossible, research has uncovered other documented cases of persons being unable to swim because their bones apparently were too heavy. In his autobiography, *I'll Tell You When You're Good!*, David Walker, an American football quarterback for Texas A&M University in the mid-1970s, said, "I can't swim a lick and never could. My bones are too heavy to even allow me to float on water. I will just sink." Another case involved a young Minnesota hockey player named George Pelawa. At the 1986 NHL draft, where he was drafted by the Calgary Flames, he told reporters "I sink in water . . . I can't float because my bones are too heavy."

An interesting footnote concerning Charlie's supposed "heavy bones" is contained in this anecdote related by grandnephew Frasier Bellam: "(My bother) Harry was a pallbearer at Uncle Charlie's funeral, and I remember him remarking that it felt like there were concrete slabs in the casket, since it was much heavier than he expected, given Uncle Charlie's slight build."

Charlie the Athlete

Artist rendition of Charlie walking underwater (Sketch by James Harvey)

OHIXIHO

Chapter 7   Charlie the Fitness Fanatic

Charlie Fraser was serious about fitness -- **very** serious. In fact, some would say he was obsessed with it. In today's vernacular, he might be called a "fitness freak." But such a description would be not only disrespectful but also inaccurate, because Charlie viewed fitness not as a fad, but as a necessity. His concern for and promotion of physical fitness long pre-dated such programs as Participaction, the Canadian government program introduced in the 1970s. Like Herman "Jackrabbit" Johannsen, the great Norwegian-Canadian cross-country skier of his generation who lived to be 112 years old, Charlie recognized that exercise was one of the keys to longevity. I clearly remember, as a boy growing up, Charlie's frequent reminder: "Whatever you do, Boy, don't forget to 'ex-a-cise'!"

Charlie's daily exercise regime consisted of a variety of elements – calisthenics, running, walking and biking. His home was equipped with exercise apparatus he designed and built himself. Consisting of a system of ropes and pulleys attached to overhead beams, this equipment helped him to stay in shape, even in his later years. He also made daily use of a small oval track that he laid out in the field behind his house. Grandnephew Harry Bellam describes the track: "When I mowed his lawn I had to also mow a path around the perimeter of his back field not more than 24 inches wide. It had to be that width so that he could jog around it, but with his poor eyesight it was easier for him to see the path if the grass was overgrown on both sides."

According to Maplemount associate Dr. Robert Paulette, Charlie continued to exercise daily to prevent the arthritis in his legs from impairing his ability to walk. Cousin Charles W.K. Fraser recalls: "He was always exercising and slept with his window wide open, in both summer and winter."

Cousin Marilyn (Fraser) Reed, who occasionally did Charlie's grocery shopping, recalls that he combined his rigorous exercise with healthy eating. "His menu consisted mainly of cheddar cheese, corned beef, bananas, figs, bread, milk, eggs, cereal and spearmint Chiclets."

As children, my siblings and I were frequently witnesses to demonstrations of Charlie's great physical condition. June (Fraser) Patterson recalls: "I remember his amazing strength. He would lie down on the floor on his back and hold his hands out for us kids to sit on them. Then he would effortlessly lift us up and down together, over and over again." Stevens Fraser remembers something similar: "Charlie would sit on a chair with his leg extended out straight and be able to support a person sitting on his leg for several minutes." As a young adult, I was often called on by Charlie to test the strength of his legs. The procedure consisted of sitting on chairs facing each other with the goal being that I use my knees to pry his knees apart. Needless to say, octogenarian Charlie prevailed, easily winning this test of strength!

Artist rendition of Charlie testing his leg strength (Sketch by James Harvey)

Running and walking were important elements of Charlie's physical fitness program. These components are covered in more detail in the previous chapter (Charlie the Athlete).

Charlie was well known for the long walks he would take, even as he increased in years. Niece Gloria (Frasier) Bellam recalled a particular incident that occurred when he was 90. "One day, while teaching in Lennoxville, I was coming home (to Cookshire) and saw him walking along the highway. I stopped to give him a ride but he said 'No' and would have nothing to do with it. When I got home, I

Charlie the Fitness Fanatic

Notes re: running and walking

telephoned (cousin) Kenneth (Fraser) to tell him about it. Kenneth drove out to pick him up, but once again Charlie refused a ride. Afterwards I talked to him about it and he said he wanted to see if he could do the walk now that he was 90. He did it and said it would not have taken him so long if so many people had not stopped to offer him a ride!"

Charlie walked everywhere. As far as is known, he never owned a car. And, as noted above, he rarely accepted a lift from anyone. When at home, he would always walk to get his groceries or to go to an auction in the next town. When travelling to Montreal or Toronto or the USA, he would go on foot to his final destination once he arrived at the bus depot, the train station or the airport, as the case may be. No taxis for Charlie!

Another vehicle Charlie used for honing his fitness was his bicycle. In 1908 he purchased a Ranger bike that was the Rolls-Royce of bicycles at that time. It had wooden rims, no chain guard and no mud guard and was the first to be equipped with New Departure safety brakes. To this he added various accessories including lights, radio, speedometer and odometer.

Charlie rode his Ranger bike for 56 consecutive years, during which time he clocked many thousands of miles biking through the rolling hills of Quebec's Eastern Townships and beyond. Cousin Stevens Fraser remembers that Charlie

Charlie walking down a street in Montreal in 1941; walking home with groceries, circa 1965 (Photos courtesy of Gloria (Frasier) Bellam)

could cycle up the very steep Fraser Hill from bottom to top, sitting on the seat the whole way. As shown in the topographical maps there is a 129-ft (39 m) difference in elevation from the bottom to the top of Fraser Hill, representing an overall gradient of approximately 10%. Some of the steeper portions probably have inclines of up to 15% -- a monumental challenge on a single-speed bike, as I well know from personal experience. Only after many failed attempts did I succeed in conquering the hill. But that was by standing on the pedals, not sitting on the seat as Charlie did.

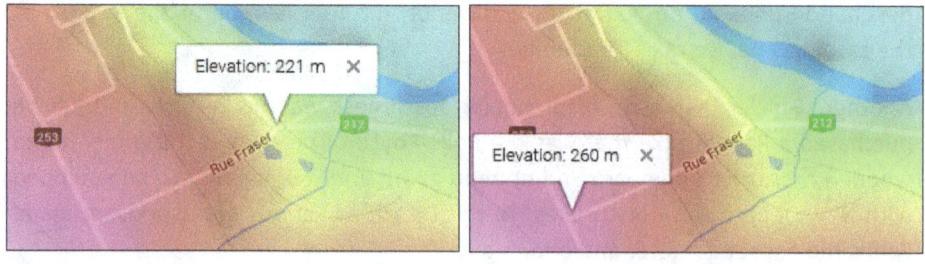

Fraser Hill elevations (Google topographical map of Cookshire, Que.)

Today, Charlie's classic machine is proudly displayed in the Compton County Historical Society Museum in Eaton, Quebec. Despite being equipped with an

Charlie's 1908 Ranger bicycle (Photo by author)

New Departure bicycle brake booklet (front, rear covers)

odometer, the exact number of miles the bike travelled is unknown simply due to the passage of time. According to Sharon Moore, president of the Museum, "I have looked at the odometer and asked a couple of other folks to take a look but we are unable to read what is below the very clouded cover."

In addition to walking, running and biking, Charlie exercised through other means. He was a member of the Sherbrooke Young Men's Christian Association (YMCA) and the Cookshire Eastview Tennis Club.

One of the amazing aspects of Charlie's fitness was how he maintained a constant weight practically his whole life. From 1911 to 1973, he kept a detailed record of his weight variation from year to year. The record includes some interesting additional notes such as "Aug. 11th, 1947: loss 8 lbs. in one day brush cutting."

Charlie's obsession with fitness might have had its roots in an incident that occurred early in his adult life. Cousin Charles W.K. Fraser explains: "One reason he walked so much was that when he was in his early 30s, he got shot with buckshot while fishing on the last day of fishing season and before hunting season started. This happened in Farnsworth's woods between Cookshire and East Angus. The doctors told him that he would never be right or well again but he showed them!"

Although he seemed always to be in good shape, Charlie had a number of medical issues that caused him to be hospitalized several times over the years, especially when he was older. It is known that he was briefly hospitalized in California during a visit there in the early 1960s. Cousin Warren Fraser, through a detailed research of his mother Alice Fraser's daily diaries, discovered that Charlie had two major surgeries in 1964, another in 1968 and yet another in 1970. It is also believed that he underwent surgery in Montreal in 1973. Because he was such a private and secretive person, almost nothing is known of the details of his operations. However, what is known are the names of the doctors and surgeons who treated him. Charlie compiled a list of medical practitioners who cared for him in one way or another during his lifetime. The list, entitled "Sons of Hippocrates," includes the names of a number of well-known doctors such as Dr. Wilder Penfield (pioneering neurosurgeon), Dr. Stanley Martin Banfill (World War II hero) and Dr. William Henry Drummond (also a poet of "The Habitant" fame).

Following his hospitalizations between 1968 and 1970, Charlie, in his unique style, published formal thank-you notes in the Sherbrooke Daily Record newspaper.

In the latter part of his life, Charlie had eyesight issues that frustrated him greatly, as illustrated in a 1968 letter to my family in which he writes "Looked for a piece of paper for 20 minutes – sight not improving. . . I get balled up very easily . – N.G. (No Good)."

YMCA and Eastview Tennis Club membership cards (Courtesy of Gloria (Frasier) Bellam)

Charles Fraser weight variation record (1911-1973)

```
              SONS OF HIPPOCRATES.
   A FEW DOCTORS placebo-ing C.C.FRASER:  1880 to 2000 A.D.


DR. HOPKINS SR.        DR. ELVIDGE          DR. GORDON
    HOPKINS A.             CRIPPS               BAKER
    DEWAR                  MCCAULEY             BABIN
    HAMILTON               MACMILLAN            MCCONNELL
    WEST                   MACREA               ORR ALF.
    PHILLIMORE             LEPINE               DRUMMOND,/"HABITANT"
    JOHNSTON               WAUGH                WILDER PENFIELD
    ALLIN                  MCCURDY              COHEN
    COUTURE                STENNING             QUINTIN
    SAMPSON                MACDONALD            BENNETT
    BANFIL                 BROWNING             KLINCK
    LYNCH W.W.             HUME                 PAULETTE
    EINBINDER              BAYNE                HICKS
    ARCHIBALD              LAMBLY               DUGAN
    LEGROS                 MCMILLAN             PENNOYER
    MCNALLY                DAVIGNON             CONE
    ROGERS                 MUNDIE


   E.& O. E.
```

List of doctors who treated Charlie

Left to right: Dr. William Henry Drummond, Dr. Wilder Penfield, Dr. Stanley Martin Banfill (Canadian Encyclopedia of Biography; McGill University Archives; McGill University Archives)

In spite of his poor vision and lameness, Charlie managed to avoid suffering a fall. Grandnephew Frasier Bellam recalls, "His balance seemed precarious, and he often walked with arms outstretched, seeming to be on the verge of falling…but he never did!" Although his vision was severely limited, Charlie could still sometimes see surprisingly well, as recounted by grandnephew Harry Bellam: "When he came over for Sunday dinner, with a patch over one eye, he could still see that Mom had put yellow margarine on the table instead of butter!"

> **ATTENTION! BEHOLD!**
>
> FRASER, C. C. — FRIENDS:: On the eve New Year wish to express publicly my deep appreciation for the many Greetings of Good Will, Gifts, Acts of Kindliness from Good SAMARITANS, Neighbors, and the General Public.
>
> May the C. C. C., Institutions, all other Societies and Organizations which contributed so generously to my WELFARE, Accept This "THANK YOU" Personally.
>
> FURTHERMORE: Including The Sherbrooke Hospital, Physicians, Surgeons, Nurses, and Staff, Directors and Supervisors of "MAPLEMOUNT".
>
> May the NEW YEAR bestow HEALTH and PROSPERITY on everybody. God Bless You All on this crazy EARTH without stint.
>
> "Toti Emul Esto."
>
> Sincerely,
> Fraser, C. C., OHIXIHO, N.A. 1968 A.D.

> **OYEZ? OYEZ? OYEZ?**
>
> **Greetings To ALL:**
>
> The Medical Staff of the Sherbrooke Hospital; Renowned Surgeons and Physicians, with the latest techniques. Members of the very efficient Outdoor Clinic, Supervisors and their spry and jolly nurses; The unexcelled Culinary Service Fare which helps one to be WELL and Upright, — A Valuable Citizen; Near and Distant Relatives, Friends, Neighbors, and the General Public, who hesitates not when needed to lend both hands.
>
> To All connected with "MAPLEMOUNT", The HOME for Young Folk; Directors, Supervisors and The Children themselves. I wish each one to consider My Heartfelt THANK YOU, A Personal One.
>
> Sincerely,
> Charles C. Fraser
> OHIXIHO
> 1969-1970
>
> Peace to the World, and
> GOODWILL to all MANKIND

Thank you notes following 1968 and 1969-1970 hospital stays

For most of his life, Charlie suffered from a severe nervous problem. Cousin Marilyn (Fraser) Reed describes how this condition was manifested. "Charlie was very nervous and therefore easily frightened. Before we would go to see Charlie, we were always cautioned to give him ample warning that we were approaching. If he were startled – and it didn't take much – he'd throw whatever he had in his

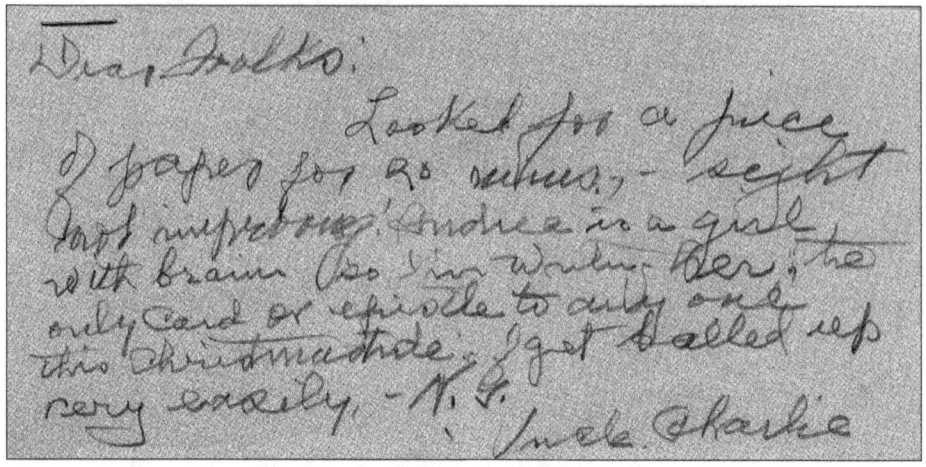
1968 letter from Charlie re: eyesight problem

hands at that moment. This often created a real danger. I remember on one occasion seeing Charlie, rake in hand, busily working on his lawn about 100 feet (it wasn't metric in those days!) away from me. I had been sent on a special mission to deliver some goodies to him (probably hot dandelion greens or delicious strawberry shortcake) since Charlie didn't cook much for himself. In order NOT to startle him, I shouted 'Hi, Charlie!' from that seemingly safe distance. Instantly the rake was tossed up in the air, his arms flew up, he jumped awkwardly, almost losing his balance and he let out a yelp. Actually, I was more frightened than he was, and also embarrassed, but I tried to smile apologetically as I offered him his dinner supplement. 'Don't scare me like that, Girl,' he admonished, as he passed his hand over his moist brow." Maplemount Home worker Ruby Campbell has a similar recollection: "He warned me never to catch him by surprise, as he would fling his arms out and hit whatever was within hitting distance!"

Chapter 8   Charlie the Chemist

Charlie's business card

Charles Clark Fraser's "day job" for more than 50 years was that of head chemist at Frasier, Thornton & Co. Ltd., a patent medicine factory in Cookshire, Quebec, that his brother Jared (Jed) was instrumental in establishing and of which his brother James (Jim) was General Manager. This enterprise was one of several where Charlie and his two brothers, Jim and Jed, worked together.

Gloria (Frasier) Bellam, Jim's daughter and Charlie's niece, describes the company's founding and early history:

> In the year 1903, at Digby, Nova Scotia, the firm of Frasier, Thornton & Co. was formed with the object in view of manufacturing and selling proprietary medicines, stove polishes and oils. Shortly after the Company was formed they purchased the Letteney Manufacturing Company, Limited of Digby , N.S., manufacturers of stove polishes and oils, also the business known as J.C. Frasier & Co., importers and manufacturers of high grade oils and greases. Within a year of being formed, the company was moved to Cookshire, Quebec, as it was more centrally located for shipping to other parts of the country, and also the town was situated on the main line of the Canadian Pacific Railway and junction of the Maine Central Railway.
>
> The year 1904 was a disastrous one for the new company, as they suffered a severe setback by having their factory, machinery and practically all their stock destroyed by fire. A new building was secured, and manufacturing soon resumed. The firm's goods were introduced and advertised in Nova Scotia, Prince Edward Island, New Brunswick and Quebec. The result of the increased and continually increasing business in this territory was that the output of the factory was inadequate to fill the orders, even when taking into consideration that the factory was running overtime about three days

Left to right: Brothers Jared (Jed), Charlie and James (Jim)
(Photo courtesy of Gloria (Frasier) Bellam)

Letteney Manufacturing Co., Ltd., Digby, N.S. (Photo courtesy of Gloria (Frasier) Bellam)

Letteney Manufacturing Co. .td,
Digby, N.S.
Manufacturers of Stove polishes & oils.
Purchased by Frasier, Thornton & Co.
in 1903.
Note- inside horns of oxen tied together
for driving them- no yoke used.

Original Frasier, Thornton & Co. building in Cookshire, Que., circa 1903 (Photo courtesy of Gloria (Frasier) Bellam)

New Frasier, Thornton & Co. building in Cookshire, Que., circa 1910 (Photo courtesy of Gloria (Frasier) Bellam)

a week. During the winter of 1907 and spring of 1908 a new building was constructed having three storeys with a floor space of 15,000 square feet. At the back was a box factory 28' by 60' where cases used in shipping goods were made. In 1911 approximately 150,000 feet of boards were converted into shipping cases. The new building was one of the first with inter-office telephones – there were about 12 installed within the facility.

The company's products were advertised in various ways. In addition to the giant ads painted on the exterior walls of the factory, there were promotional items such as blotters and rulers. They also advertised on the back of their company envelopes.

Gloria (Frasier) Bellam continues her story of the company's history:

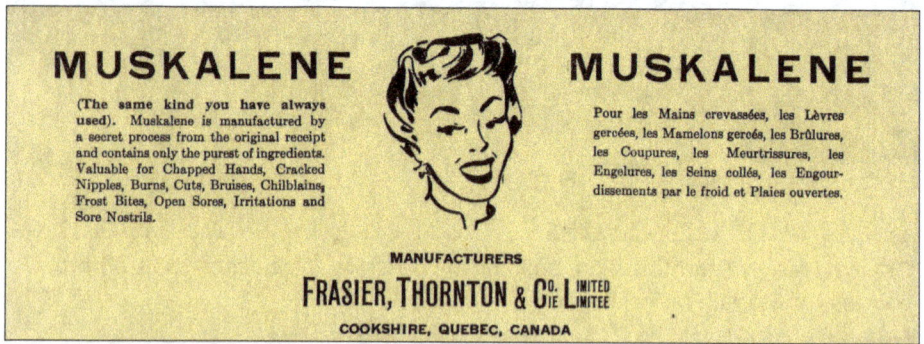

Advertising blotter for Muskalene (From the author's collection)

Oliveine Emulsion advertisement on back of an envelope (From the author's collection)

Frasier, Thornton & Co. promotional items (Photos by Bertina Benitez)

Between the years 1904 and 1912, the following companies were bought, one at a time, and their preparations added to the growing list of Frasier, Thornton & Co. products:
- Fred L. Shaffner & Co., manufacturers of Dr. J. Woodbury's Horse Liniment and Dr. J. Woodbury's Condition Powders
- The Dr. Bell Medicine Co., Truro, N.S., manufacturers of Dr. Bell's Remedies
- The Acme Mfg. Co., Lunenburg, N.S., manufacturers of Acme Stock Food Remedies

- Pendleton's Panacea Co., St. John, N.B., manufacturers of Pendleton's Panacea and Pendleton's Remedies

In addition, the company was sole owner and manufacturer of:
- Dr. Stanley's famous preparations
- Oliveine Emulsion
- Laxa Purple Quinine Tablets
- Muskalene
- Pelletier's Syrup of Tar and Codene
- Frasier's Oils and Greases
- Thornton's English Veterinary Remedies.

The photograph below shows a broad selection of the company's patent medicine products. The subsequent photo presents close-up views of such exotic remedies as Dr. Stanley's Mosquito Oil, Frasier's Syrup with Wild Cherry and White Pine Tar and Dr. Stanley's Jamaica Ginger.

Selection of Frasier, Thornton & Co. patent medicines (Photo by Bertina Benitez)

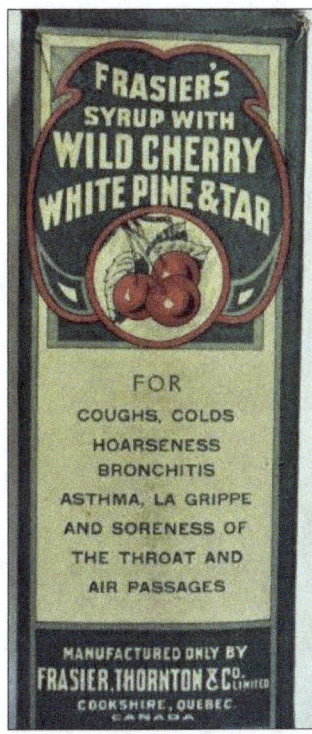  

Close-ups of Frasier, Thornton & Co. patent medicines (Photos by Bertina Benitez)

Also included in the company's product line were less exotic preparations such as cod liver oil.

As head chemist, Charlie was intimately involved in the development of the formulae for these various medicines and other preparations. The fact that a common ingredient of many of the medicines was alcohol caused the company some unforeseen problems. Gloria (Frasier) Bellam provides the details:

> The western branch (of Frasier, Thornton & Co.) in Saskatoon (Saskatchewan) opened in 1910 and flourished with sales increasing rapidly over the next few years. However, when

Cod liver oil product card

James (Charlie's older brother) heard why some of the preparations, which contained alcohol as a preservative, were being bought in such large quantities he was very concerned. People were extracting the alcohol from the cough syrups etc. for drinking. Because James' (and Charlie's) mother had been an ardent W.C.T.U. worker and had signed a pledge never to take alcoholic beverages, he closed the western branch of the company, and from then on, sales were chiefly in Quebec and the Maritimes where alcoholic beverages were more easily obtained.

Several years later, the company faced another alcohol-related challenge. Around 1942, during World War II, the federal government gave the breweries an increased allotment of alcohol, but an increase was refused to Frasier, Thornton & Co. Ltd. However, after company representatives went to Ottawa to protest, they returned with an increased allotment.

Charlie not only developed the recipes for the company's patent medicines but also coined the names for several of them, including Oliveine Emulsion and Muskalene. These two product names survived for the life of the company, but

Artist rendition of Frasier, Thornton & Co. protesting in Ottawa (Sketch by James Harvey)

not all of them did. The federal government forced the company to change the name of "Pendelton's Panacea" to "Pendelton's **Pancea**." To protest the change, the company modified the label to print as shown opposite, by retaining the middle "a" but putting an oblique bar through it.

Pendelton's Panácea

Modified label for Panacea (Courtesy of Gloria (Frasier) Bellam)

The government also tried to make the company change "Dr. Stanley's White Liniment" to "Dr. Stanley's Eggless Liniment." But the company refused and continued with the original name for that very popular product.

Although Frasier, Thornton and Co. was best known for its patent medicines to treat human conditions, it also offered various veterinary products such as horse liniment and condition powder. In addition, the company sold non-medicinal products such as harness oil.

The company's complete product line is shown in an undated price list on the following pages.

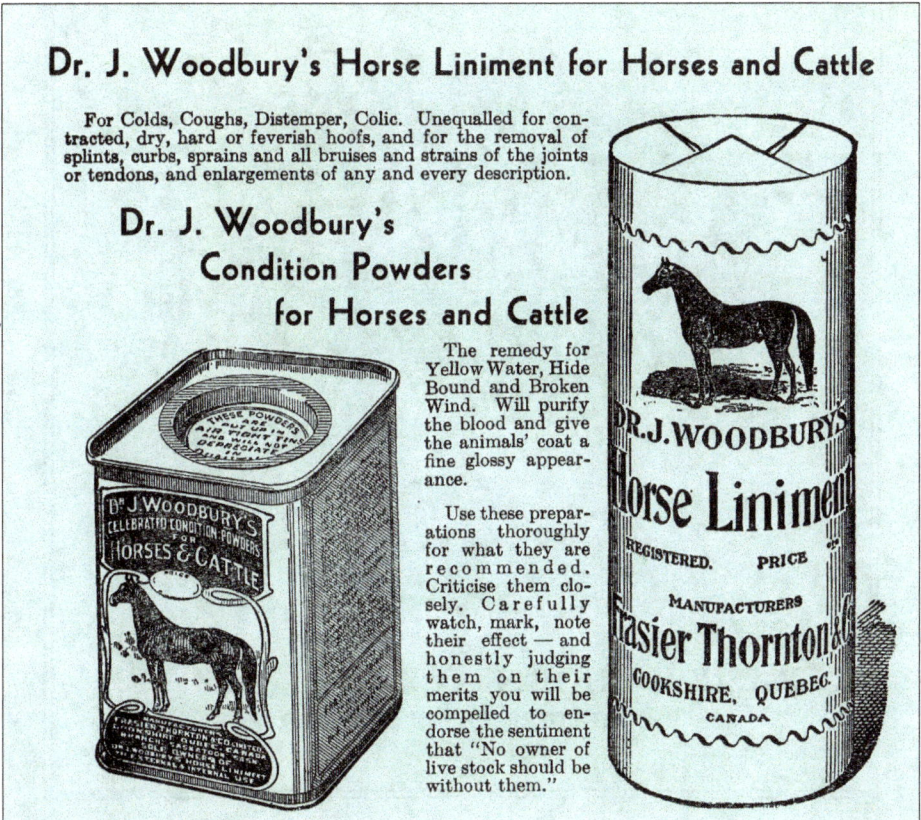

Veterinary products

FRASIER, THORNTON & CO., LIMITED

COOKSHIRE, QUEBEC, CANADA

THE LARGEST PROPRIETARY MEDICINE COMPANY IN CANADA
MANUFACTURING AND SELLING EXCLUSIVELY GOODS
OF THEIR OWN MANUFACTURE

PRICE LIST

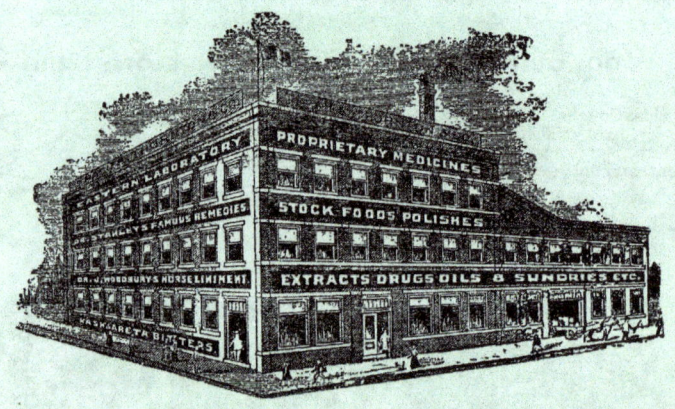

Manufacturers of
PROPRIETARY MEDICINES, TOILET PREPARATIONS,
VETERINARY REMEDIES AND SUNDRIES

FRASIER, THORNTON & CO., LIMITED
COOKSHIRE, QUEBEC, CANADA

Frasier, Thornton & Co. product price list, front cover

DR. STANLEY'S

	PER DOZ.
AFRICAN PINE	4.85
BABY OIL	6.25
BELLADONNA PLASTERS	4.85
CARBOLATED SALVE	3.85
COD LIVER OIL	6.25
COD LIVER OIL (LARGE)	12.00
CORN SALVE	2.75
FOOT POWDER	3.25
JAMAICA GINGER	4.85
MOSQUITO OIL	4.65
PAIN EASE RUB	4.65
POWERFUL WHITE LINIMENT	3.85
ROOT PILLS	5.25
STOMACH AND LIVER TABLETS	4.85
SWEET CASTOR OIL	4.25
TOOTHACHE REMEDY	4.25
TALCUM POWDER, BABY BORATED	3.25

DR. J. WOODBURY'S

CONDITION POWDERS	6.50
HORSE LINIMENT	6.00
GUMBO, A REVIVER AND PICK UP	10.50

ACME

HEAVE POWDER	6.75
LOUSE KILLER	6.25
POULTRY FOOD (PACKAGES)	6.25
POULTRY FOOD (35 LB. DRUMS) PER LB.	.22
POULTRY FOOD (100 LBS.) "	.22
STOCK FOOD (PACKAGES)	6.25
STOCK FOOD (35 LB. DRUMS) PER LB.	.22
STOCK FOOD (100 LBS.) "	.22
VETERINARY HEALING OINTMENT	4.35

FRASIER & THORNTON

COW HEALTH POWDER	9.00
COW HEALTH POWDER (65 LB. DRUMS) PER LB.	.35
COW HEALTH POWDER (100 LBS.) PER LB.	.35
HOG HEALTH (PACKAGE)	8.50
HOG HEALTH POWDER (65 LB. DRUMS) PER LB.	.22
HOG HEALTH POWDER (100 LBS.) PER LB.	.22
LIVE STOCK TONIC (PACKAGES)	8.50
LIVE STOCK TONIC (65 LB. DRUMS) PER LB.	.22
LIVE STOCK TONIC (100 LBS.) PER LB.	.22
LIQUID GALL REMEDY (SMALL SIZE)	4.75
LIQUID GALL REMEDY (LARGE SIZE)	9.00
VETERINARY BLISTERING LIQUID	12.00
COW BAG COMFORT	6.75
URINAL REMEDY POWDER	6.75
DUSTING POWDER	4.25
LIVE STOCK LAXATIVE AND PURGATIVE MIXTURE	9.00

No sales tax charged as a separate item—No charge for packing cases and cartons—Free goods for full amount of transportation.

Frasier, Thornton & Co. product price list, p.2

	PER DOZ.
FRASIER'S	
EXTRACT OF WILD ROOTS	4.85
FAMOUS RED CASTOR HARVESTING OIL	6.25
FURNITURE POLISH	3.50
HANDY FAMILY OIL	3.50
RELIABLE CREAM SEPARATOR OIL	8.25
SYRUP WITH WILD CHERRY, WHITE PINE AND TAR	4.85
OLIVEINE	
EMULSION	10.50
OINTMENT	4.50
PILLS	6.00
THORNTON'S	
CARBOLIC HEALING OIL	4.65
GALL REMEDY PASTE	4.25
PAIN REMOVER	4.65
SYRUP WITH WHITE PINE AND TAR	4.25
WORM SYRUP	6.25
GOLDEN LION BRAND	
BORAX	1.50
EPSOM SALTS	1.35
SENNA	1.35
SULPHUR	1.35
HELP-A-COLD LAXA TABLETS	4.85
HUGHES COLIC REMEDY	12.00
KUMFORT RUB (ALCOHOL ISOPROPYL)	7.25
KUMFORT RUB (ALCOHOL ISOPROPYL), SMALL	4.00
KASKARETA BITTERS	4.85
MUSKALENE	5.25
PELLETIER'S SYRUP WITH OIL OF PINE TAR & MENTHOL	4.85
PENDLETON'S PANCEA	6.00
PINE PITCH KIDNEY PILLS	4.85
SEMINOLE BABY COUGH SYRUP	4.50
SEMINOLE LIQUID LOUSE KILLER	4.65
SEMINOLE OIL LINIMENT	5.75
ACETYL SALICYLIC ACID TABLETS	3.25
CAMPHORATED OIL (U.S.P.)	5.45
COD OIL FEEDING .... PER GALLON	
COMET METAL POLISH	4.35
EXCELSIOR LIGHTER FLUID	3.25
FLY-U-DI HOUSEHOLD SPRAY 5% D.D.T.	4.00
GERM-U-DI (LARGE BOTTLES)	9.00
GERM-U-DI (BOTTLES)	3.50
GERM-U-DI (1 GAL. CANS) .... EACH	
OIL OF TAR	3.50
RAT KILL	4.25
SEWING MACHINE OIL	2.75
SODA MINT TABLETS	2.50
SPIRITS NITRE (B.P.)	4.50
TINCTURE IODINE (1 OZ.) (B.P.)	2.75
SPOT OUT	3.50
PAROL (PURE MEDICINAL MINERAL OIL)	8.25
TURPENTINE (3 OZS.)	3.25
BENZINE	3.25
BORACIC ACID	2.65
DIXIELAND PINE TAR	4.85

Bonus goods on 3 dozen or more of the same article.—Liberal cash discount allowed according to amount of purchase.

Frasier, Thornton & Co. product price list, p.3

As chief chemist, Charlie was keeper of the secret recipes for the company's various preparations. His extensive raw materials list (see Appendix D) contains more than 100 items including the following less than appetizing ingredients:

- white pine bark
- squill root
- chloroform
- turpentine
- charcoal
- kerosene
- isopropyl
- ethyl nitrite
- varsol

Charlie shopped far and wide for these ingredients, as evidenced by a postcard dated 1919 from J. L. Hopkins & Co. of New York, suppliers of roots, barks, herbs, leaves and seeds.

Although their products were marketed only in Canada, a number of Frasier, Thornton & Co.'s suppliers were from the USA. These included Whitney & Kemmerer (coal supplier) of Buffalo, N.Y.; the Buckeye Stamping Co. of Columbus, Ohio; the Chemical Supply Co. of Cleveland, Ohio; Synfleur Scientific Laboratories of Monticello, N.Y.; and United Naval Stores Co. of New York, N.Y. as indicated by mailed items.

Harness oil product card and advertisement

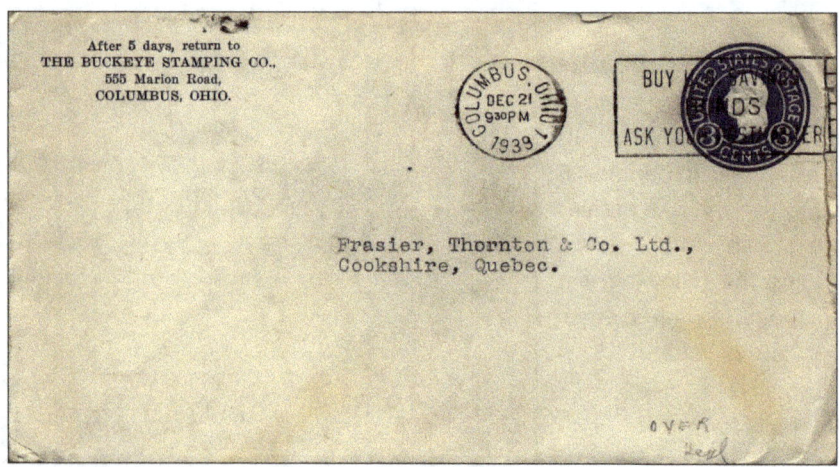

Envelope from supplier Buckeye Stamping Co.

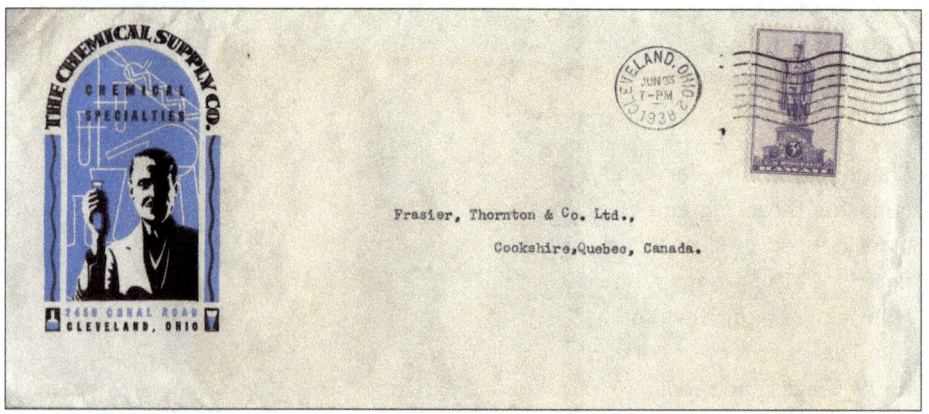

Envelope from supplier Chemical Supply Co.

Envelope from supplier United Naval Stores

Charlie the Chemist

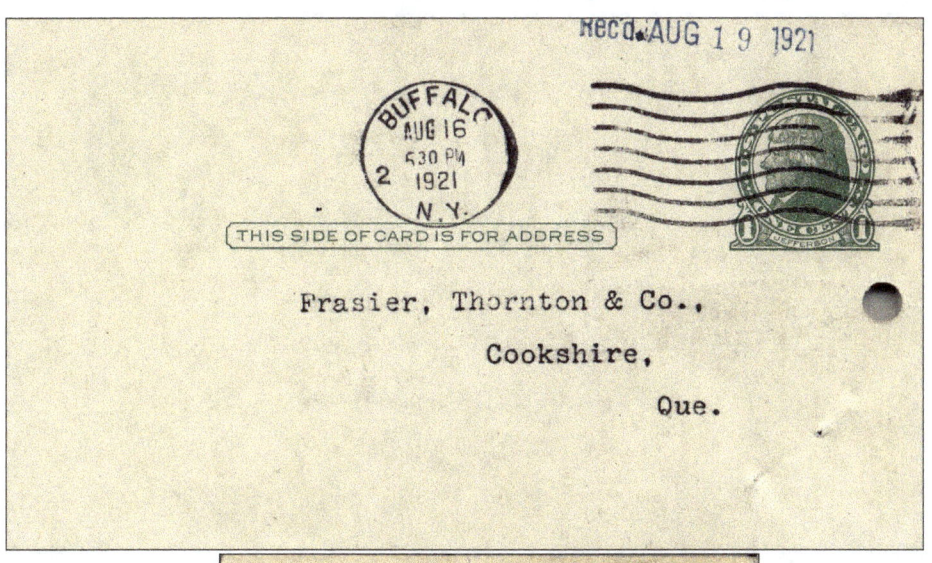

Front and back of postcard from supplier Whitney & Kemmerer

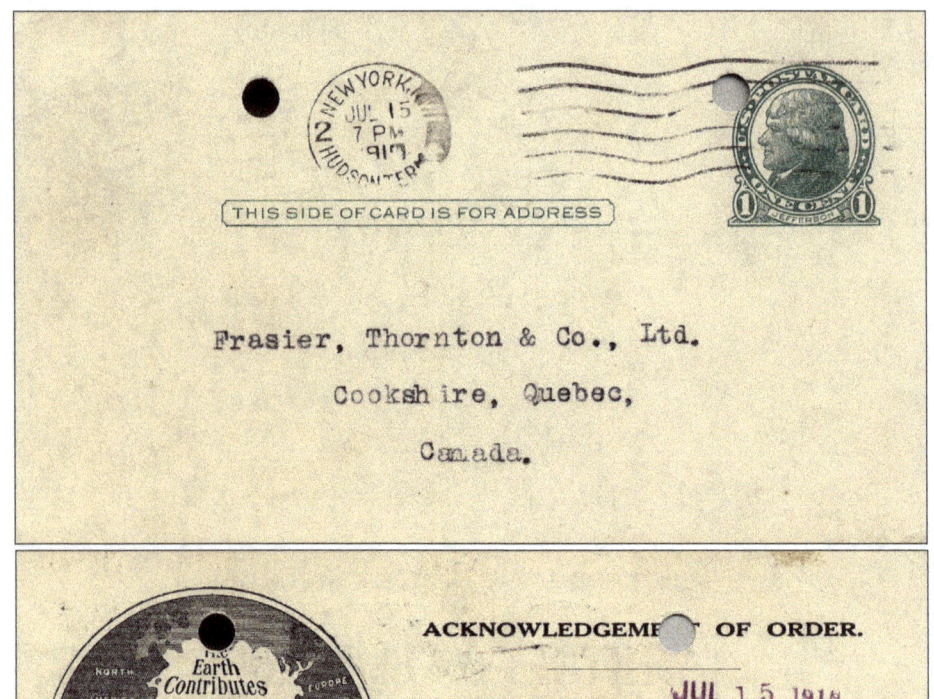

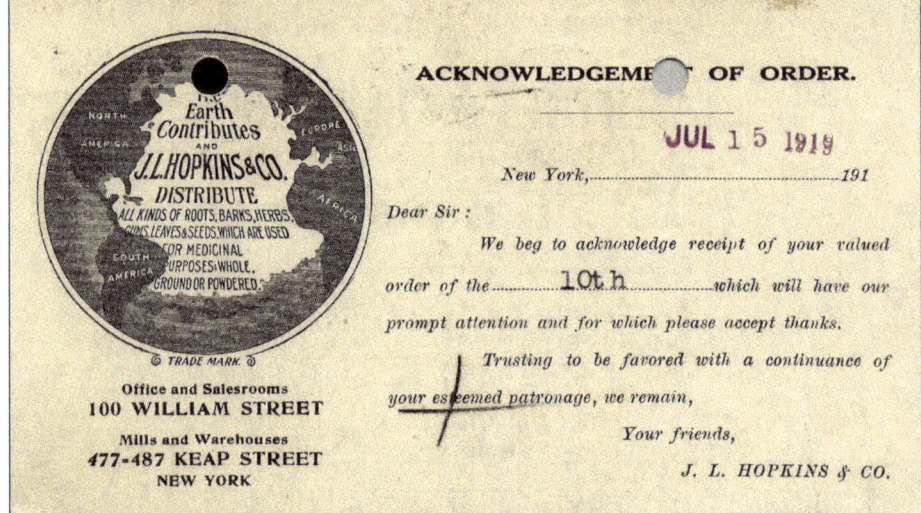

Front and back of postcard from supplier J. L. Hopkins & Co.

For more than 50 years, Frasier Thornton & Company was a respected institution and its advertisement-covered building a unique landmark seen by thousands, thanks to its prominent location beside the main line of the Canadian Pacific Railway from Montreal to Halifax, N.S. During that half-century, Charles Clark Fraser was a key contributor to the company's success.

During its first two decades of operation the company's growth was quite phenomenal. Unfortunately this was not sustained in the later years of its history. In her 2003 University of Waterloo master's thesis "The Fortunes and Failures of Frasier, Thornton and Company 1903-1969: A History of a Proprietary Medicine Manufacturer in its Canadian Context," Julie Carolyn Mayrand writes:

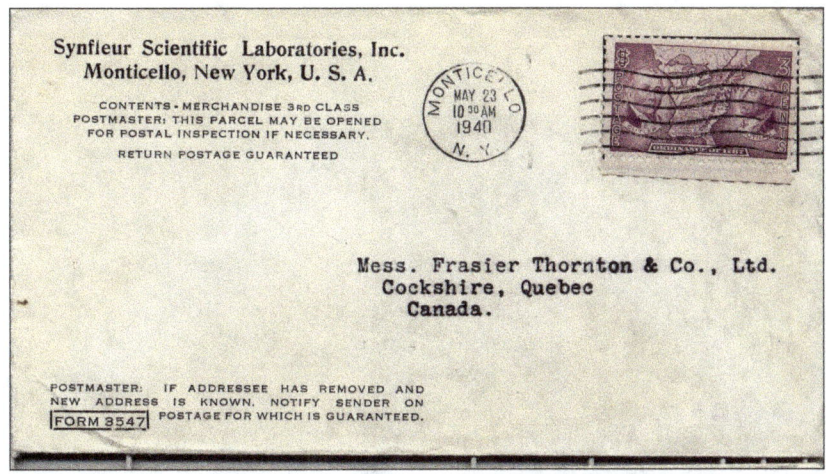

Envelope from supplier Synfleur Scientific Laboratories, Inc.

Sales reached $127,853 in 1919 and climbed to $137,117 in 1920. Through these years of limited government regulation, Frasier and Thornton had managed to make a relative success of their little venture. However, the legislative changes brought in by the government in 1919, however, (sic) marked the beginning of a long period of decline for the firm. To claim that Frasier, Thornton and Company's downfall was the introduction of legislative controls in Canada would be far too simplistic an explanation. Though admittedly it did not quite manage to recover financially after 1927, the company managed to survive another fifty years after the 1919 amendments to the Proprietary or Patent Medicine Act were brought into effect. It was the combination of legislation, poor economic conditions during the inter-war period, and changing attitudes towards medicine that hampered the company's growth. As late as the 1930s, proprietary remedies still dominated the market; there were few effective alternatives as modern medicine was still in its infancy. Therefore, it was not until the late 1940s and early 1950s that modern pharmaceuticals, such as penicillin, began outselling the old fashioned remedies. Competing with options that actually cured, it is no surprise that businesses like Frasier, Thornton and Company began to close their doors.

Although the company folded in 1969 and its ad-plastered building was demolished, the legacy of this unique manufacturing facility lives on. Cousin Jim Fraser, editor of the Fraser Family Link newsletter, notes in the July 2008 issue: "Their bottles have become collector's items and are regularly offered on eBay. . . Frasier, Thornton and Co. bottles are important symbols of our medical history and are displayed in museums across Canada and (beyond)."

Two Frasier, Thornton & Co. proprietary medicine embossed bottles. Left: Bottle offered for sale on eBay (Photo from eBay listing); right: Pendelton's Panacea bottle purchased on eBay by Art Pease (Photo by Jim Fraser)

Steam engines on the C.P.R. main line, Cookshire, Que., circa 1950 (Photo by author)

C.P.R. station, Cookshire (Photo from eBay listing)

OHIXIHO

Chapter 9   Charlie the Collector

Charlie Fraser was a collector extraordinaire. He was simultaneously a coin collector, stamp collector and collector of curios. For him, these activities were not half-baked hobbies or part-time passions. Collecting was a serious vocation to which he dedicated much of his time, energy and resources.

Numismatist

It is significant to note that on the very first line of his autobiographical summary, Charlie identified himself as "Numismatist & Philatelist." He was obviously proud to describe himself in this way. And well he should have been. He was known across the country as a major collector of early Canadian coins and paper money. Charlie was a very early member of the Canadian Numismatic Association (CNA). His membership number was No. 205 of an association that grew to more than 10,000 members by 1973. He attended a number of annual CNA conventions where he would add to his collection through purchases from or trades with other collectors.

Little is known of the details of his collection, but cousin Charles W. K. Fraser recalls a special set of paper notes that Charlie owned. "He once had the five lowest denomination bills ($1, $2, $5, $10 and $20) all with the same serial number (although the prefix letters were different). My dad was certain that Charlie must have had friends at the Mint!" Whether or not he had friends at the Canadian Mint remains a question, but he **did** have friends who helped him build his collection. Among them was Howard Barter, local bank manager, who was a collector himself. Mr. Barter would always keep an eye open for any rare coins or paper money that passed through his bank.

As Charlie's sight began to fail when we was in his eighties, he sometimes solicited my help to decipher the almost invisible dates on some of his oldest coins. How exciting that was for a youngster who had never seen such pieces of money!

Family members were often beneficiaries of Charlie's interest in coins and paper money. Niece Gloria (Frasier) Bellam remembers that every Christmas her Uncle Charlie would give her a "shinplaster." A shinplaster was the common name for the 25-cent Canadian paper bill issued from 1870 to 1935.

Cousin Marilyn (Fraser) Reed has similar memories: "I remember receiving my first silver dollars from Charlie as gifts for various occasions. To this day, I still have them, as I was given the impression many years ago that these were meant to be

Shinplaster 25-cent note

kept and not to be spent."

Being a coin collector, Charlie always had a large cache of surplus pennies. Instead of wrapping them and taking them to the bank, he had a more creative way to dispose of them. True to his generosity of character, he devised a unique way to give them away. Every Christmas (and occasionally at other times)

Canada Silver Dollar, 1960

he would arrive at his cousin Donald Fraser's home with a pouch full of pennies for the "Fraser 12" children. Everyone would gather in the living room and each child would be assigned a specific date or range of dates. One by one, Charlie would pull a penny from his pouch and call out its date. The lucky kid who was assigned that date would receive the penny. By the time the game was finished, amazingly every child ended up with the same amount of money. One of those children was the late Marina (Fraser) Tracy, who recalled in an article in the 1993 Fraser Family LINK newsletter: "It made us feel rich!"

According to cousin John (Jack) Fraser, in the mid-1960s Charlie sold his entire collection to Canada's leading coin dealer, James E. Charlton, based in Toronto. Charlton, who lived to be 102 years old, was also a numismatic publisher whose price guides represented the authoritative standard for all serious collectors. True to Charlie's secretive nature, details of the sales transaction were known only to himself.

Philatelist

Charlie was an avid stamp collector who belonged to a number of philatelic associations and attended numerous conventions and exhibitions. For example, he was a supporting member of the Canadian Association of Philatelic Exhibitions and attended the Canadian Centenary International Philatelic Exhibition in 1951 in Toronto.

Charlie was a Life member of the Pioneer Philatelic Phalanx (PPP), an exclusive association of philatelists that was founded around 1900. He also served the group as its Canadian Ambassador of Goodwill. Internet research reveals the characteristics of the association: "The PPP had no dues, no executive, and no magazine. It met at different shows across the U.S. To be a member you had to have been a (stamp) collector in the 19th Century." (Reference: http://www.bnaps.org/hhl/newsletters/mm/mm-2001-05-n150.pdf.) One of the group's annual meetings is covered in the November 1940 issue of Hobbies magazine. In

CAPE membership card

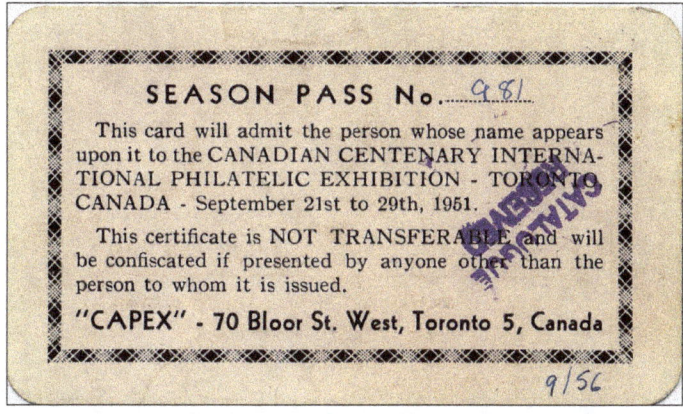

CAPEX pass

the article Charlie's name appears in the company of some very important persons of that era, including President Franklin Roosevelt.

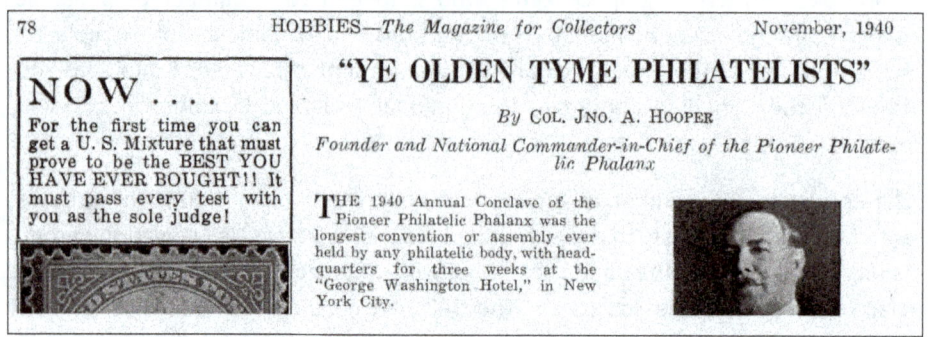

Coverage of PPP meeting in Hobbies magazine

> Letters of congratulations received by the Phalanx included ones from President F. D. Roosevelt, former Postmaster General Farley, Hon. Frank Murphy, of the U. S. Justice Department, Ramsey S. Black, Asst., P.M.G., and from our Canadian ambassador-of-good-will, the Hon. Chas. C. Fraser.

Reference to Charlie in Hobbies magazine article

In his search for stamps to enhance his growing collection, Charlie dealt with suppliers and collectors across the United States, as evidenced by the selection of covers (envelopes) following.

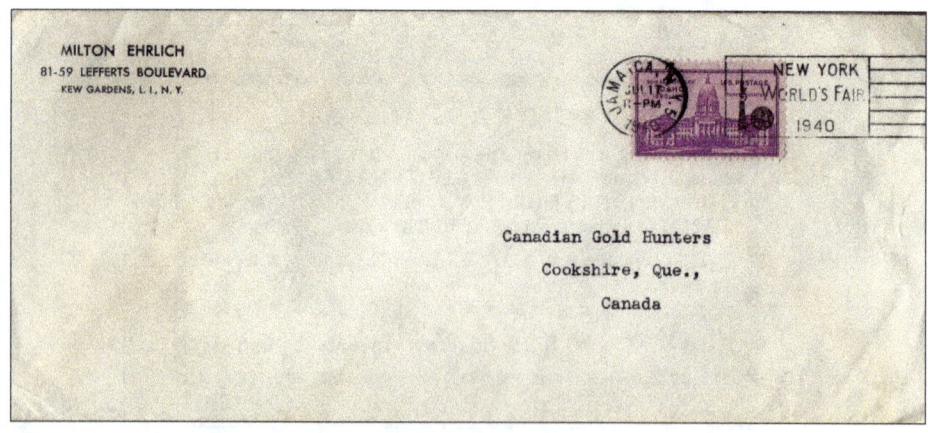

Letter from Milton Ehrlich, Kew Gardens N.Y.

Letter from Elmer R Long, Harrisburg, Penn.

Letter from Central Iowa Stamp Co., Marshalltown, Iowa

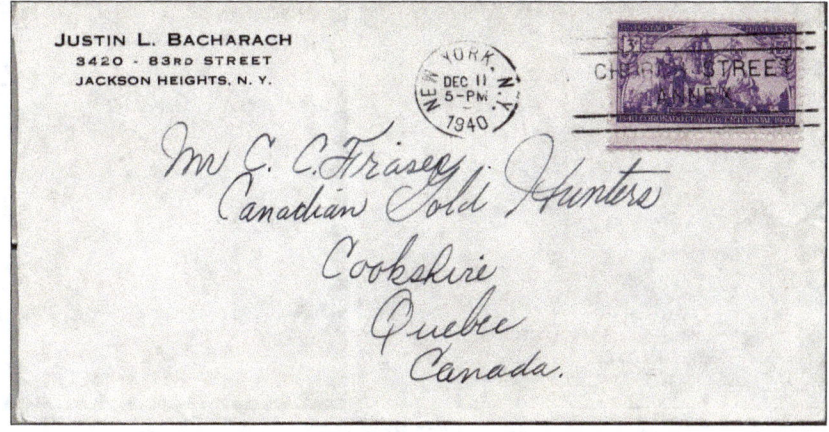

Letter from Justin L. Bacharach, Jackson Heights, N.Y.

An interesting handwritten note appears on an envelope date-stamped 1938 (see figure below). The note reads: "meaning of cap on "2" 1890-93 U.S.A."

Envelope with "cap on 2" note

This mysterious note was explained by an Internet search. It refers to a flawed issue of a Washington two-cent stamp that seems to show a baseball cap on top of one or both 2s (see figures below).

> The reason for the "Cap on the left 2 and Cap on both 2s" occurred during the transfer of the image from the transfer relief plate to the soft plate that was used for the printing. These breaks occurred when the transfer relief plate was not properly hardened. Due to stresses tiny stress breaks resulted

Washington 2-cent stamp with cap on "2" (Photos from http://www.stampcommunity.org/topic.asp?TOPIC_ID=6289 and ID=23084)

and were missed during the final inspection prior to the transfer of the image to the final soft plate.
(Reference: http://www. stampcommunity.org/topic.asp?TOPIC_ID=6289.)

It is not known whether Charlie actually had this rare stamp in his collection.

It appears that one of Charlie's philatelic specialties was "first day covers" since his collection contained a wide variety of these valuable collectibles. "A first day of issue cover or first day cover (FDC) is a postage stamp on a cover, postal card or stamped envelope franked on the first day the issue is authorized for use within the country or territory of the stamp-issuing authority." (Reference: https://en.wikipedia.org/wiki/First_day_of_issue.)

Following is a selection of Charlie's first day covers, including a tongue-in-cheek cover issued in Eden, Ill. E. Koestler was one of the more prolific cover producers with whom Charlie communicated.

Among the postal oddities in Charlie's collection was a 1927 air mail letter that was stamped as "LETTER RETURNED" due to "AIR FLIGHT ABANDONED." It is noted that air mail service was in its infancy at that time. Only in the following year, 1928, did the Canadian Post Office begin providing official Air Mail services.

Five-cent Air Mail First Day Cover

Half-cent First Day Cover

Lyndon B. Johnson Inauguration First Day Cover

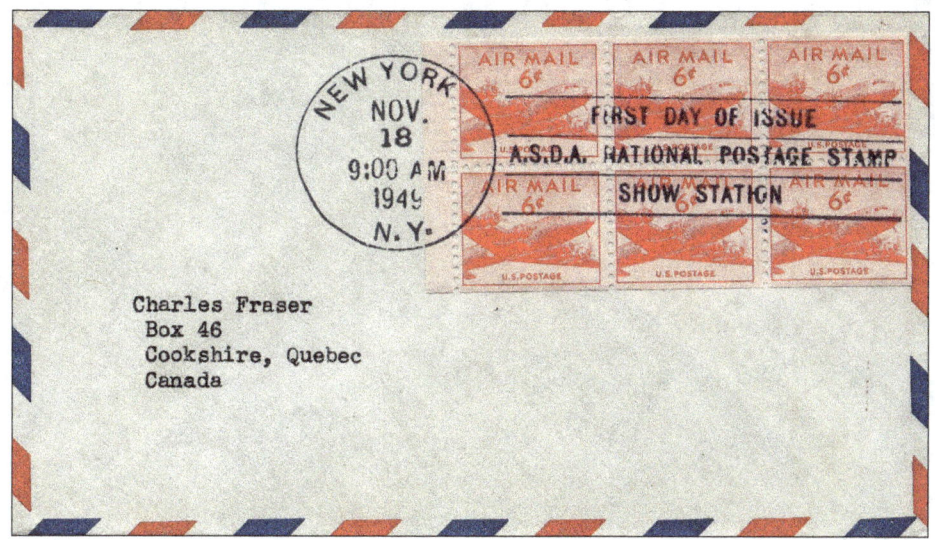

A.S.D.A. First Day Cover

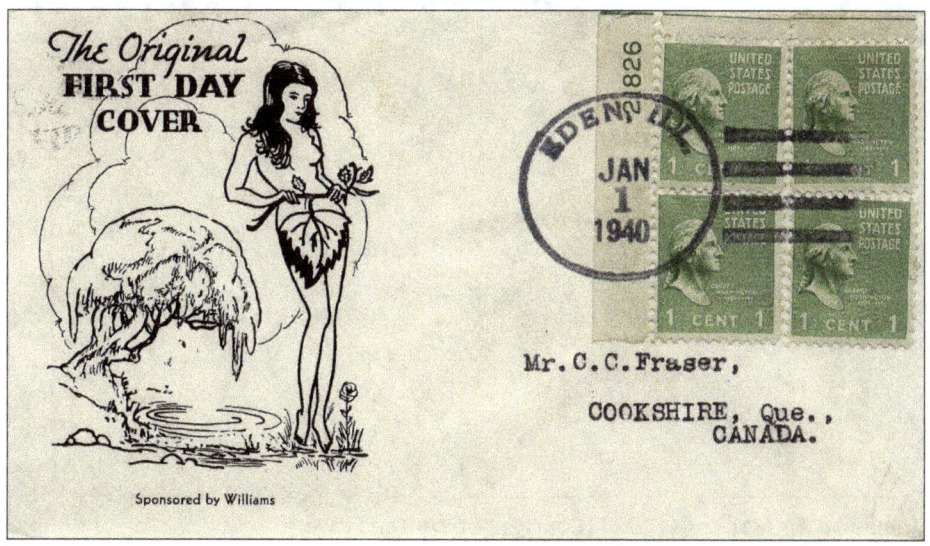

"Original" First Day Cover

Letter from cover producer A. E. Koestler

Canadian Air Mail cover 1927 (Courtesy of Warren Fraser)

Curios Collector

Charlie was a compulsive collector of every kind of curio. Some came from his prospecting adventures in northern Quebec. Others were of more local provenance. Cousin Marilyn (Fraser) Reed tells the story of one of his curio hunting experiences:

> Charlie loved auction sales and frequently walked miles to collect "treasures." One day, he had been especially successful at a sale in Eaton Corner or Sawyerville, but was too late in realizing he would not be able to carry all the new acquisitions in his arms. He solved the problem by buying an old-fashioned baby carriage, piling in his purchases and pushing the loaded buggy home to Cookshire. He delighted in the amused glances from passers-by along the highway.

Artist rendition of Charlie transporting auction treasures in baby carriage (Sketch by James Harvey)

When Charlie moved from his family home, Maplemount, into his "new" home, OHIXIHO, in 1958, he turned the entire second floor into a museum. The new house was formerly the Maplemount carriage house that was moved on beams and rails to its new location.

Moving Maplemount carriage house onto OHIXIHO foundation (Photo by author)

The OHIXIHO museum was filled (perhaps "cluttered" would be a better word) with Charlie's accumulated collection of curiosities. Among the most memorable items were:

- The first western moving picture travelling outfit (in which a man swallows a mouse!)
- A piece of the rope that was used to hang a man in the Yukon
- An antique set of bailiff's handcuffs complete with the key
- A pair of Indian "medicine" trees
- A genuine Irish shillelagh
- An antique human shoulder yoke
- Antique sewing machines
- Gramophones

Grandnephew Harry Bellam was especially intrigued by one of the items he was shown: "I was invited into the house and shown a few things, and I always wondered why somebody would keep glass X-rays of himself with buckshot from the time he got shot in the buttocks!"

OHIXIHO house and museum, 1972 (Photo by Jim Fraser)

OHIXIHO detail (Photo by Warren Fraser)

Most of these items were ticketed with tags containing neatly-typed information pertaining to their origin. For example, the tags for the vintage handcuffs (pictured below) are shown opposite.

The OHIXIHO Museum was open to the public and admission was free. Upon entry, a visitor would be issued a "Concert" ticket with the date handwritten on the back. Then Charlie would usher the visitor to the steep stairway and pull a rope to open the overhead trap door to the second floor. Once upstairs, the guided tour would begin.

According to niece Gloria (Frasier) Bellam, Charlie attempted unsuccessfully to have the Quebec Government declare OHIXIHO an official museum. Nevertheless, he continued for many years to educate and entertain the local public through his one-of-a-kind personal museum. When Charlie had to leave OHIXIHO due to failing health, many of the museum's items were donated to the Compton County Historical Society Museum in Eaton, Quebec, an institution he helped to found in 1959, and the very same building where he was baptized in 1884 (see Appendix C).

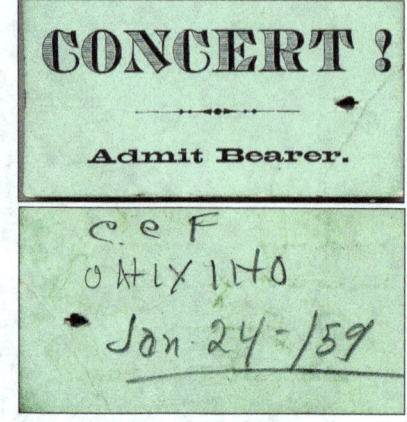

Front and back of OHIXIHO Museum admission ticket

According to museum president Sharon Moore, among the numbered items logged in the museum's donations book for 1975 were the following from Charlie's collection:

- 616 Radio
- 621 Bricks (made from local material)
- 623 Gramophone
- 630 Log roller
- 631 Cant dog
- 632 Viewer
- 634 Yoke
- 635 Typewriter
- 644 Blacksmith vise
- 649 Bicycle
- 650 Lamp
- 651 Lamp wick holder
- 652 Scrapbook of old calling cards or greeting cards

Cookshire bailiff William Wilford's handcuffs, circa 1880 (Photo by author)

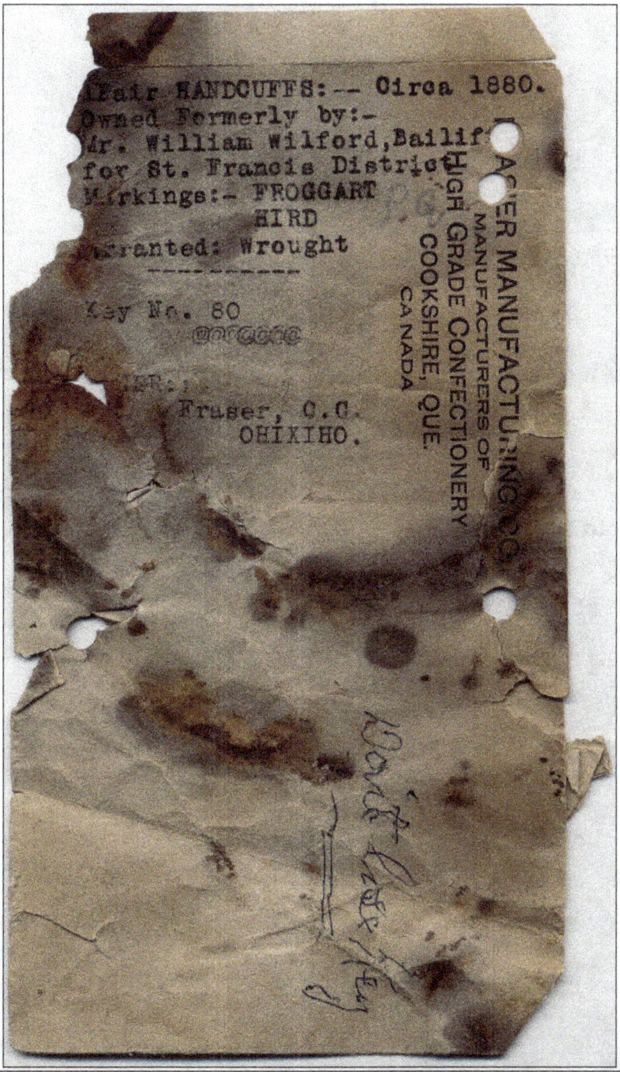

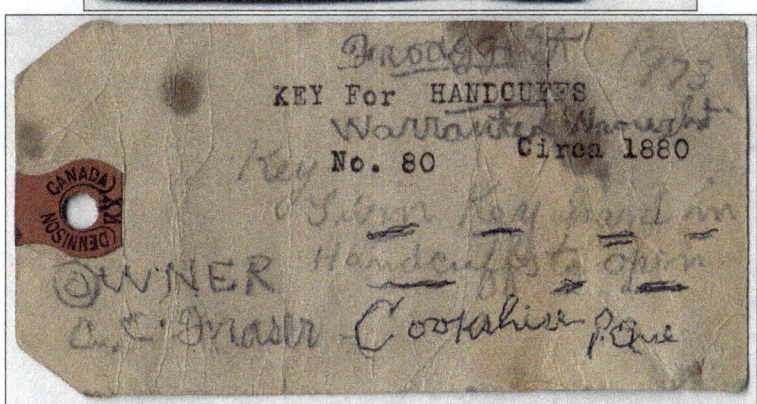

Item information tags for antique handcuffs

Human shoulder yoke (Photo by author)

Shillelagh (Photo by author)

"Medicine tree" (Photo by Carolyn Byerlay)

Compton County Historical Society Museum, Eaton, Que. (Photo by author)

Chapter 10   Charlie the Jack of Many Trades

On his autobiographical summary sheet (see Appendix A), Charlie described himself as "Jack of All and One Trades." In addition to the vocations already covered in the preceding chapters, he mentions the following occupations:
- Confectionery Manufacturer
- Fox Rancher
- Woodsman
- Carpenter
- Farmer
- Stationary (Steam) Engineer
- Trapper
- Hunter
- Fisherman

Other trades or occupations that he did not mention but was known to have had were:
- Photographer
- Animal Trainer
- Blacksmith
- … and who knows how many more!

Charlie was also a "would-be" soldier. According to niece Gloria (Frasier) Bellam: "Uncle Charlie tried to join the Army to fight in World War I but was refused, probably for health reasons because he had recently been shot while fishing by a hunter who mistook him for a bear."

This chapter will limit its coverage to only a few of the above -- those for which at least a basic amount of detail is known.

Confectionery Manufacturer

Charlie was involved in a candy factory in Cookshire (Quebec) started by his older brother Jed in 1918. The factory was located on Railroad Street not far from the Frasier, Thornton & Co. factory.

Niece Gloria (Frasier) Bellam recounts the history of the short-lived Frasier Manufacturing Company:

> After collaborating with Mr. Thornton to establish Frasier, Thornton & Co. in 1903, by 1918 Jared had had so many disagreements with his business

partner that he left the company, and started up a candy factory. He engaged a candy maker from Sweden who knew how to make all types of hard candy, many with designs in the centres. The glucose for the candy came in 700-800 lb. hogsheads (kegs). Their candy was sold in 30-pound pails.

Fraser Manufacturing Co. sign (Photo courtesy of Sharon Peart Waite and Sally Aldinger)

They made only hard candy. As there seemed to be a demand also for chocolate bars, Jared ordered a train carload of chocolate bars from Willard's in Toronto. These sold so quickly that Jared ordered a second carload before the due date of the payment for the first carload. Willard's sent their Credit Manager to Cookshire to check up on the candy factory. As a result of the visit, a second carload was dispatched. It was a good thing that the bars sold quickly, because the weight of a whole carload of them made the building begin to sag! The business flourished until 1923 when Jared's (and Charlie's) mother died. Jed was so traumatized by this event that he closed his factory.

Among the candy factory mementos is an envelope from the company addressed

Candy factory building as it looked in 1985 (Photo courtesy of Gloria (Frasier) Bellam)

Willard's Chocolate Factory in Toronto, 1919 (From http://hockeygods.com/images/10875Willard_s_Chocolate_Factory_453_Wellington_West_Toronto_1919)

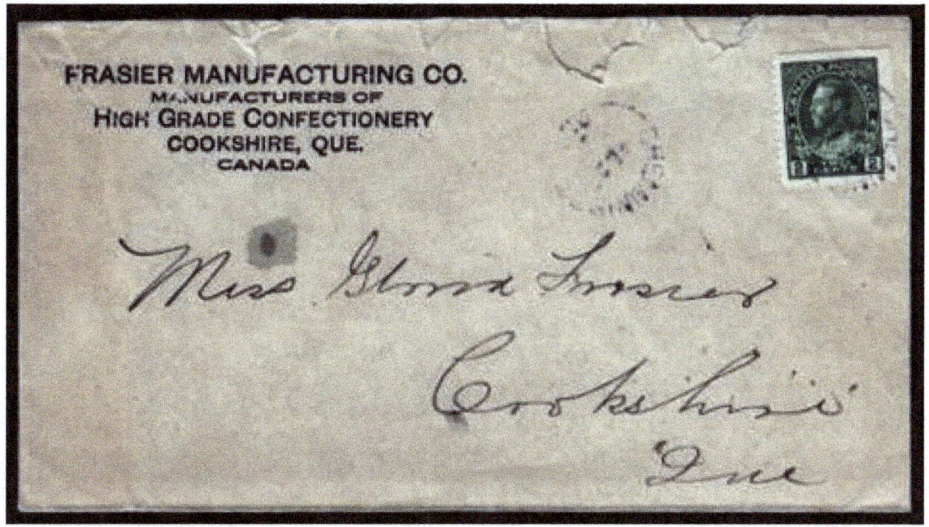

Envelope addressed to Gloria Frasier from Frasier Mfg. Co. (From author's collection)

to Charlie's niece Gloria.

In addition to hard candy and chocolates, the company also sold salted peanuts in small pouches made of the recently invented cellophane material.

Fox Rancher

The reader may be surprised to learn that Charlie was a fox rancher. In fact, around 1910, he and his brother Jed established what was only the second fox ranch in the province of Quebec. The purpose of fox ranches was to raise the beautiful silver fox, which was prized for its pelts, especially in Europe.

Frasier's salted peanuts bag

The fox ranching industry in Canada has a very interesting history, as illustrated by some excerpts from a 2004 Legion Magazine article "When Fox was King" (reference https://legionmagazine.com/en/2004/09/when-fox-was-king/):

> Those who got in at the right time made fortunes, literally overnight. And it all came from farming—farming with a furry twist. The name of the game was fox farming or ranching, and for several years it remained the hottest industry on Prince Edward Island. The late 19th and early 20th centuries brought hard economic times to many Islanders. . . . The development of fox farming at the turn of the century brought a much-needed, but

unfortunately short-lived economic boom to P.E.I., and from 1900 to 1940 the Island was the world leader in this new industry.

Although the foxes were the offspring of native red animals, they were black with a sheen of silver in their fur, as their outer guard hairs were tipped with it. They were one of the rarest animals in nature and much prized in a society where being rich meant

Silver fox (Photo from www.arkinspace.com -Image Credit Flickr User Matt Knoth)

wearing fur. Known variously as silver, black, black silver and silver black, the fur trade used the term "silver fox" to cover all grades, from light silver to pure black. . . .

Up until 1910, (six breeders) completely controlled the province's domestic fox industry and, by extension, the world's. But word did get out that fortunes could be made from foxes. Prices on the international market climbed to unheard of heights . . . One prize skin alone fetched over $2,600. This was excellent money considering that the average annual wage for an Island farm labourer at the time was roughly $225, while a university professor earned $1,200. . . . Fox fur was much in demand by the nobility, especially Russian and Austrian. With Island fur so sought after, prices soared, especially for prime pelts. . .

(In the) 1930s, the demand for fox pelts declined dramatically. When money was scarce, fashion took a back seat to more pressing concerns. . . . Between 1938 and 1941, fox farming started to fall into insignificance and

Frasier's fox ranch in Cookshire, Que., circa 1914 (Photo courtesy of Gloria (Frasier) Bellam)

by the end of the decade pelt prices reached an all-time low as fashion tastes changed to lighter coloured and more exotic furs. Many farmers simply opened their pens and let their once-valuable animals escape into the woods.

Frasier's fox ranch was located in a field behind the family home. It was described in a brief article that appeared in the Sherbrooke Daily Record in 1914.

According to niece Gloria, Jed and Charlie's fox ranch was quite short-lived. "The ranch continued for 6-8 years, but after the foxes were all stolen, they closed down the ranch." It is not clear how profitable the venture was but it must have been economically worthwhile in order to have lasted as long as it did.

> **FOX RANCH AT COOKSHIRE**
>
> THERE ARE TWENTY ANIMALS IN ENCLOSURES OWNED BY FRASIER BROS.
>
> Cookshire, Oct. 28 —(Special)— There are, perhaps, very few towns of the size of Cookshire which can boast of a thriving fox ranch such as is owned by Frasier Bros. This ranch situated on a high elevation of land adjoining their property, is fenced off into enclosures with wire netting, the size of each enslosure being 32 feet square. In each of these is a den with a passageway leading to it. At present 22 foxes, comprising 11 pairs are housed here, and among this number are several beautiful animals. Mr. J. C. Frasier, one of the proprietors has signified his intention of not allowing visitors to the ranch after the 1st prox., and while there are a few days open to visitors it would be well worth the while of all interested to call on Mr. Frasier and ask him to show them around.

Sherbrooke Daily Record article on Frasier fox ranch

Woodsman

Charlie was an experienced and expert woodsman. He wielded the broad-axe and the crosscut saw with equal dexterity. In 1958, when he needed some 30-foot-long beams for his "new" house to rest on, he went to his cousin Donald's woodlot with broad-axe in hand and quickly created the required lumber.

Charlie's woodsman experience started when he was very much younger. Niece Gloria tells the story: "My father (James Frasier) and Uncle Charles were engaged by some man to go to his woods and cut wood all winter. They would get paid in the spring. So they cut wood all winter and come spring, the man would not pay them. He said that there was nothing written down as an agreement, so they did not get paid!"

This was not Charlie's only negative experience as a woodsman. It is believed that he lost the index finger of his left hand while operating a circular saw powered by a steam engine such as the one illustrated.

Charlie and cousin Donald Fraser hauling squared logs, Cookshire, Que., 1958 (Photo by author)

Charlie unloading squared logs, Cookshire, Que., 1958 (Photo by author)

Sawing wood with circular saw powered by steam engine (© Shaun Dibley, Alamy.com)

Animal Trainer

Pigs are known more for wallowing in the mud than for their intelligence. But people like Barnum & Bailey and Charles Clark Fraser have proven that pigs are smart animals that can be trained. An 1895 poster for Barnum and Bailey's Circus features "A Remarkable Troupe of Trained Pigs" that could play music.

Although obviously not in the same league as the circus people of his day, Charlie was nevertheless a successful animal trainer. What he did was quite remarkable -- he trained his pet pig to sit at the table and hold a coffee cup between its hoofs! Unfortunately, no photos of this spectacle have been

Muddy pigs (Photo by author)

Barnum & Bailey poster, 1895 (Reference: http://pigofknowledge. pi-blogspot.ca/2006/12/circus-pigs.html)

discovered, but a number of people vouch for having witnessed the sipping swine's performance.

Artist rendition of Charlie's pet pig (Sketch by James Harvey)

Photographer

Although Charlie took hundreds, perhaps thousands, of photographs of others, he hated having his own picture taken. He rarely allowed anyone to "shoot his mug," so most photos that do exist were probably taken without his knowledge or permission. Cousin Warren Fraser's spontaneous portrait of Charlie in his later years is such an example.

Given Charlie's strong aversion to having his picture taken, it was surprising to find among his papers a two-sided photograph of himself (i.e., two photos glued together, back to back). One side was a full length portrait facing the camera and the other side was of similar composition but with his back to the camera. It is unknown who took these pictures. Could it have been Charlie himself by using a tripod and timer? If so, then he could be credited with having invented the "selfie" many decades before it became the fad that it now is!

Charlie unposed (Photo by Warren Fraser)

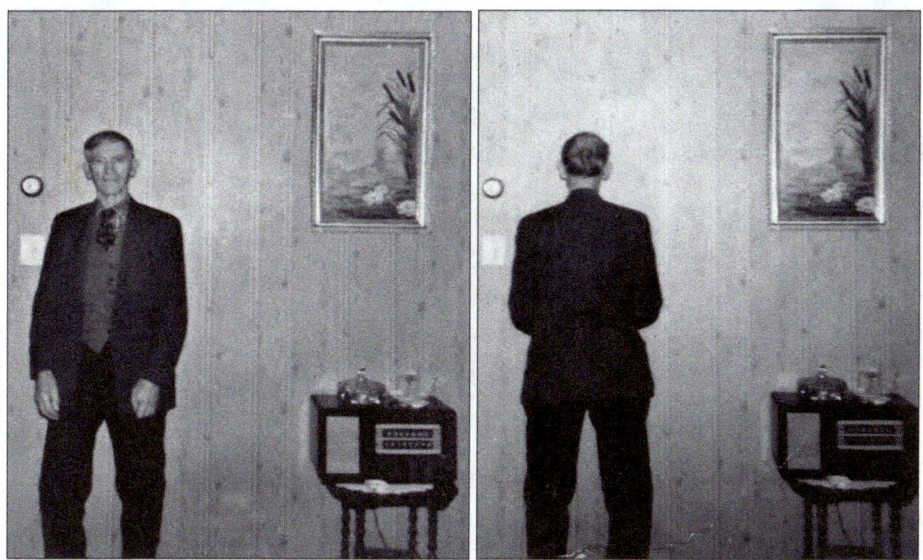

Charlie self-portrait, front and back (Photos courtesy of Gloria (Frasier) Bellam)

Carpenter

Charlie's competence as a carpenter was demonstrated when, with the help of an assistant, he completely gutted the interior of a barn and transformed it into a house.

Demolition of barn interior (Photo by author)

Lumber stored in barn for OHIXIHO house construction (Photo by author)

Chapter 11   Charlie the Philanthropist

An Internet search provides a number of different definitions for philanthropist. Among them are:

- a person who seeks to promote the welfare of others, especially by the generous donation of money to good causes
- a person giving time or valuables (money, security, property) for public purposes
- one who makes an active effort to promote human welfare

Regardless of the definition selected, Charles Clark Fraser would most definitely qualify as a bona fide philanthropist. His life was punctuated with acts of generosity, many of them known only to him and the recipient. Some were more public in nature. For example, he was a Life Member of the Canadian Bible Society, indicative of being a major financial supporter.

But Charlie's crowning action as a philanthropist occurred when he was 82 years old. A few years earlier he had moved out of his large family homestead, Maplemount, into a smaller house nearby. The old house stood vacant for a few years while Charlie considered what to do with it. But following conversations with his physician, Dr. William Klinck, he decided to donate Maplemount to become a home for young folk. As cousin Charles W. K. Fraser recalls: "Charlie sold Maplemount for one dollar." Thus was born the Maplemount Project.

Dr. Klinck's Sherbrooke Hospital associate, Dr. Robert Paulette, recounts how he became involved in the project. "It was in early 1962 that I was informed of the gift of Maplemount . . . This enterprise was engineered by Dr. W.J. Klinck, a graduate of Western Medical School in London, Ont., who set up his practice in Lennoxville in 1939. In his practice, that included Cookshire, Charlie was a patient of his and offered to convert Maplemount into a home for young folk. Since Dr. Klinck and I worked together at the Sherbrooke Hospital and worshipped together at Grace Chapel, I was enlisted. As secretary-treasurer, I acknowledged all the donations that got the Home going."

An interesting side note is that Dr. Klinck was made a Member of the Order of Canada in 1997. The accompanying citation mentioned Maplemount Home.

In the months that followed Charlie's generous donation, a Board of Directors was established, planning meetings were held and a major renovation of the building was undertaken. The renovated Maplemount had a fresh new look as compared to the stately but drab old Maplemount.

Old Maplemount (Photo courtesy of Charles W.K. Fraser)

New Maplemount (Photo by author)

Cars lined up for public meeting at Maplemount, 1963

Order of Canada

William Klinck, C.M., M.D.C.M.

Full Name	Honour Received	Residence
William Klinck, C.M., M.D.C.M.	C.M.	Lennoxville
Honour	**Appointment**	**Investiture**
Member of the Order of Canada	April 17, 1997	October 22, 1997

This Lennoxville surgeon has given compassionate medical relief to his patients in Canada and in Third World countries. As a volunteer missionary, he has travelled at his own expense to some of the world's poorest nations to teach and heal. A true humanitarian, he has lessened the suffering of some of the most vulnerable members of society, including battered women and children, the homeless or those struggling with substance abuse. He played a pivotal role in the establishment and management of two local institutions dedicated to the care of senior citizens, in addition to founding Maplemount Homes for children in need.

Above: Dr. Klinck's Order of Canada Citation, 1997
Inset: Dr. William Klinck (Photo courtesy of Dale White)

Ruby Jean Campbell, who worked at Maplemount from 1963 to 1968, and is currently a missionary in Colombia (South America), recalls her experience:

> My main job was doing the cooking but also caring for the children, dressing the little ones, putting them to bed, etc. . . . The most children we ever had at one time was 31. Twenty-six was actually our limit but (one day) this French-Canadian family arrived without notice, left their children and then took off – leaving no forwarding address! I forget how many weeks passed before (the children) were "reclaimed"!

Thanks to "Uncle" Charlie's generosity and the dedication of workers like Ruby, many children from difficult home situations were beneficiaries of Maplemount Home's loving and caring environment during a critical period in their lives. One of the children, Tom, now working in the mission field himself, remembers his time at Maplemount:

> I am 100% open to commend the Maplemount project. The great love of people like the Schmidts, Munkittricks, Dr. William Klinck and many others cannot be forgotten. I wish they'd been more appreciated. My years at Maplemount were most of the best years of my life as a child. As for our cook, "Aunty Ruby," we loved her. I don't remember much about the food except that there was always lots, always delicious, and we loved meal time! The best reason for being grateful for Maplemount is that it was there that I came to know the Lord!

According to Dr. Robert Paulette, "Mr. Fraser took a very active interest in the welfare of the kids at Maplemount." Living just a stone's throw away, Charlie was a frequent presence at the Home where he chatted with the children and staff alike.

Because of Charlie's and Dr. Klinck's visionary initiative, many others were motivated to acts of charity.

During its operation through the 1960s and early 1970s, the Home was well supported by the community at large through donations of money, food, clothing, quilts, toys and various other items. The figure below shows some newspaper

Hatley Center W.I. makes donation to Maple Mount Home	Cookshire	Bishopton
	A meeting of the St. Margaret's Guild and W.A. of St. Peter's Church, was held at the home of Mrs. Gilbert Ross. $15.00 was voted to the Maplemount Home towards the Dental Clinic work.	The Happy Gang Club met at the home of Mrs. George Latewood last Wednesday when a quilt was finished for the Maplemount Home in Cookshire

Local newspaper clippings re: Donations to Maplemount Home

clippings of such donations.

In the mid-1970s, the work that had been taking place on the Maplemount property in Cookshire was decentralized to three individual homes in Huntingville. Community support and involvement continued. For example, the "BCS Bulletin, November 1974" mentions that "Eight students travel weekly to Huntingville to tutor children at the Maplemount Home." The organization overseeing the Homes' operations, Maplemount Homes Inc., was dissolved in 1988, thus ending a quarter-century of service to young people – that all began because of Charlie's benevolence.

Chapter 12   Charlie the Man

Above and beyond his many accomplishments, Charlie was a family man, a man of character and a man of mystery. When asked what adjectives come to mind to describe him, various persons who knew him have offered the following: a gentleman, eccentric, unusual, unique, fascinating, intelligent, brilliant, secretive, stubborn, kind, caring, thoughtful, delightful, compassionate and helpful. California cousin Sharon Peart Waite, who met him when she was very young, described him simply as "a nice old fellow."

First and foremost, Charlie was a family man. This may seem to be rather a strange characterization of someone who was a lifelong bachelor (or as he expressed it in his autobiographical summary, "confirmed and confounded bachelor evermore"). Found among his collection of clippings was the poem A Bachelor's Vow.

In spite of apparently subscribing to the views expressed in the poem, Charlie was a man who loved being with family and doing for family. He

```
            A BACHELOR'S VOW  / 1864.
                    "           "
OH, I WILL A B A C H E L O R  BE,
NOR BIND MYSELF FAST TO A SHREW;
FOR A BACHELORSS HAPPY AND FREE,
WHILE WEDLOCK HAS BLISSES BUT FEW.

THOUGH MY SOCKS ARE OUT AT THE TOES,
AND THE BUTTONS OFF OF MY SHIRT,
NO VIXEN SHALL PULL MY POOR NOSE,
AND VOW THAT I'M COVERED WITH DIRT.

OH, I'LL NEVER, NO, NEVER SUBMIT
TO MARRY A GAILY DRESSED DOLL ,
AND PROMISE, BY ALL THAT IS "WRIT,"
I'LL PROTECT HER FROM EVERY FALL.

OH, I'LL NEVER, IN ALL MY LIFE,
TAKE A FLOUNCED AND SIMPERING FLIRT,
WHO WOULD, WHILE SHE SAT AS MY WIFE,
LEAVE THE FLOOR ALL COVERED WITH DIRT.

THEN TO THINK OF SATIN AND SILK,
AND HOOPS AND IMPORTED LACES,
OF RIBBONS AND EVERY SUCH ILK,
AND paint TO BEDAUB THEIR FACES.

THEN TO THINK OF THOSE " FILAGREED" THINGS,
NEW BONNETS, SO HUGE AND SO GRAND,
WHICH COST ,TO SAY NOTHING OF STRINGS,
AS MUCH AS AN ACRE OF LAND.

TO SEE HER AT NINE IN THE MORNING,
THE PAINT DAUBS HALF WASHED FROM HER FACE;
THE CHILDREN CRYING AND SCREAMING,
IN DRESSES TOO LOOSE, MUCH, FOR GRACE.

TO SIT AT THE BREAKFAST TABLE,
AND FIND THAT COFFEE IS COLD;
THE TOAST BURNT TO THE HUE OF SABLE,
THE BACON BOTH RANCID AND OLD.

THE CHILDREN PULLING AND HAULING,
THEIR FINGERS IN EVERY DISH;
WHILE MA, THEIR MANNERS BEWAILING,
YET GRATIFIES EVERY WISH.

AND THIS IS BUT PART OF THE GRIEF
THAT BEFALLS A " FAMILY MAN:"
TO SOME IT MAY BE PAST BELIEF,
BUT I WILL ESCAPE IF I CAN.

AGAIN, THEN, THIS OATH I'LL RENEW,
TO ALWAYS LIVE single AND free;
FOR RATHER THAN MARRY A SHREW,
I'D THROW MYSELF INTO THE SEA.    O.S.W.

OHIXIHO
1 9 6 4
C.C.F.
```

A Bachelor's Vow (Author unknown)

was never married and had no descendants, so who was his family? In fact he had several families. His first family, i.e., his immediate family, consisted of his parents

Charlie Fraser's family home, "Maplemount," Cookshire, Que. (Photo courtesy of Charles W.K. Fraser)

Gloria and Darrell Bellam with children Frasier, Liles and Harold (Harry), 1967 (Photo courtesy of Frasier Bellam)

and siblings as well as his niece Gloria (Frasier) Bellam and her family. Since most of his siblings also never married, they remained in the family home, Maplemount, their whole lives. In their later years, it was Charlie who provided for their care.

The family housekeeper, Mrs. Katie Burton, was hired around 1944 and became an important part of that family for many years. Later, during the final few years of Charlie's life, Katie cared for him in her own home on the Lennoxville-Waterville highway. Niece Gloria describes this period of her uncle's life: "When Charlie was about 96 years old, he was persuaded by Katie to go and live at her home so that she would be able to care for him. She knew how to handle him, for he was very stubborn. By this time, Katie's husband John had died, so she was able to fully cater to Charlie's needs. He died in his 98th year."

Katie Burton home, near Waterville, Que. (Google street view photo)

Charlie's second family was that of his first cousin, Donald Fraser, his wife Alice and their 12 children who lived on a farm in Cookshire less than a mile away. A very frequent farm helper and evening visitor (Alice's diary records more than 400 such instances over a three-year period in the early 1950s), it is also suspected that he provided financial assistance to this large family.

Charlie's third family was his "Maplemount Home for Young Folk" family (see Chapter 11, "Charlie the Philanthropist"). The children and their care-givers called him "Uncle Charlie." In the words of Dr. Robert Paulette, secretary-treasurer of the Home: "He loved the kids. It was as if he inherited a large family when he didn't have any children of his own."

Other groups to which Charlie belonged that might also qualify as families were the Pioneer Philatelic Phalanx (PPP, an association of stamp collectors), the Ancient Order of Petrologists (a private "lodge"), and the Cookshire Christmas Club (a women's group). The PPP is described in Chapter 9, "Charlie the Collector."

Donald and Alice Fraser and the "Fraser 12," circa 1962 (Photo by Dick Tracy)

Details of the Ancient Order of Petrologists are unknown but it is believed to have been a sort of private men's "lodge" established by Charlie and some of his closest friends. According to Wikipedia, petrology is "the branch of geology that studies the origin, composition, distribution and structure of rocks." (reference https://en.wikipedia.org/wiki/Petrology). Therefore it might have been related in some way to his prospector vocation.

According to long-time Cookshire resident Dorothy Ross, the Christmas Club was a women's group that prepared and distributed Christmas baskets to the less-privileged of the community. How Charlie became a member of this women-only group remains somewhat of a mystery, but it is strongly suspected that he was a major financial sponsor of this charitable work. The objectives of the Christmas Club are described in the 1889 Annual Report of The Church Society of the Anglican Diocese of Quebec:

The Christmas Club is an association of ladies who, for the few weeks preceding Christmas, give up two or three hours each Saturday to make the celebration of our Lord's Nativity brighter for the needy families of the Parish. Gifts for children, articles of clothing, not always new, but of value to the recipients, together with certain offerings of money, are gathered and distributed. And many expressions of gratitude for the kindness thus shown have reached the ears of the clergy.

Although the Cookshire Christmas Club disbanded several decades ago, the Christmas Club in neighbouring Bury, Que., is still active today.

Apart from his families noted above, Charlie forged a number of individual friendships, some of which lasted for years. One such example was his friendship with Patricia Stevenson Smith, whom he met when she was a young schoolgirl and with whom he maintained contact for many years well into her adult life. Pat relates how they met and how this special relationship developed:

> My earliest memories (of Charlie) are when we moved to Cookshire in June of 1959. All the kids talked about this man at OHIXIHO. I think they used to play pranks on him. I was anxious to meet him. So I introduced myself and thus began our friendship. I was probably 12 or 13 when we

A WONDERFUL MACHINE

Man is the world's most miraculous mechanism. Want proof? Look at these figures.

In 70 years of life, a human being eats 1,400 times his own body weight, more than 100 tons of food, and he spends five full years just putting food into his mouth. If his weight is average, every day his heart beats 103,680 times; he breathes 23,040 times; he inhales 438 cubic feet of air, he gives off eighty-five degrees (F) of heat; he moves 750 major muscles; and he utters approximately 4,800 different words.

The average person blinks his eyes 25 times a minute, and scientists say that each blink lasts an average of one-fifth of a second. Thus, if he averages 40 miles an hour on a ten-hour motor trip, he drives 25 miles with his eyes shut. The human body can take a lot of punishment and still function.

It's funny, though, one accident can throw in the monkey wrench that can stop this wonderful machine cold !

Clipping "A Wonderful Machine"

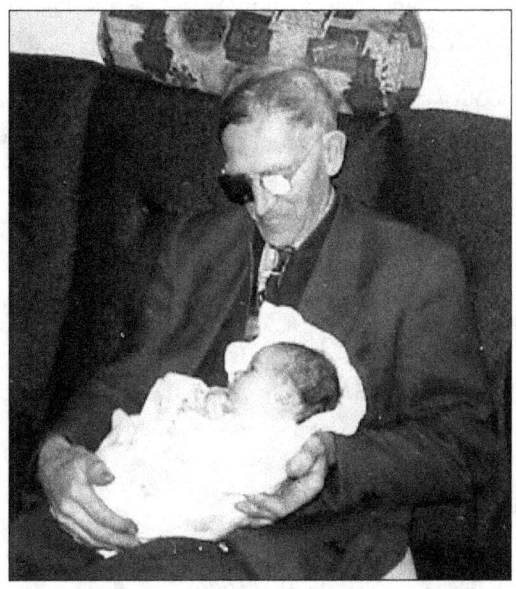

Charlie holding 4-week-old Stephanie Lynn Smith (Photo by Patricia Stevenson Smith)

first became acquainted. He was like the grandfather I never had. We shared a special bond.

We would visit quite regularly and when I was at Quebec Lodge, Macdonald College and teaching in Rosemere, we exchanged letters. We corresponded for years. I always enjoyed receiving his letters. Sometimes he typed and sometimes he wrote (by hand). Treasures for sure. (One letter contained) a piece entitled "A Wonderful Machine." Also in (another) letter he wrote a definition of sleep: "Tired Nature's Sweet Restorer." I remember making him a guest book out of birch bark at camp one summer. When our first daughter was born in 1971, he came to visit us at our apartment in Lennoxville.

As demonstrated above, Charlie took a particular interest in and had a special love for newborn babies. Just before our first child was born, he sent a letter containing his suggestion for the baby's name. A few weeks later he sent another letter, this one addressed to baby Andrea Jane, enclosing "2 smacks to be applied by Proxy."

Charlie was very dedicated to his various families. He was always there to help, especially in emergencies. As noted in his obituary (see Appendix C): "To organizations and individuals alike, he lived the motto that 'A friend in need is a friend indeed.'" Niece Gloria recalls how her Uncle Charlie came to the rescue when her mother fell outside and broke her knee in 1957. "I could not move her. Mother said 'Phone Charlie.' He immediately came over and got her onto a rug and moved her onto the verandah…. to wait for the ambulance. He was very helpful and compassionate . . . He would help anyone. He thought a lot of his family."

Cousin Malcolm Fraser recalls his dad telling about the time lightning struck their barn filled with hay and set the roof on fire. Charlie stayed up all night to keep watch to prevent the live embers of the charred beams from igniting the hay barely a foot beneath.

Charlie was very interested in family genealogy and maintained correspondence with various cousins he had tracked down. Examples of these letters show a writing style that is both formal and folksy.

Charlie kept in regular contact with his first cousin, Neva Frasier Cauzza, who lived in California. Neva loved and admired him greatly and always addressed her letters to "Hon. C. C. Fraser."

Charlie was a stubborn man who refused to give up until a particular goal was achieved, whether it was solving a mathematical problem or nursing someone back to health. In the 1960s, his farmer friend, Malcolm Maclean, stricken with cancer, was told by the surgeon that he would never walk again. Refusing to

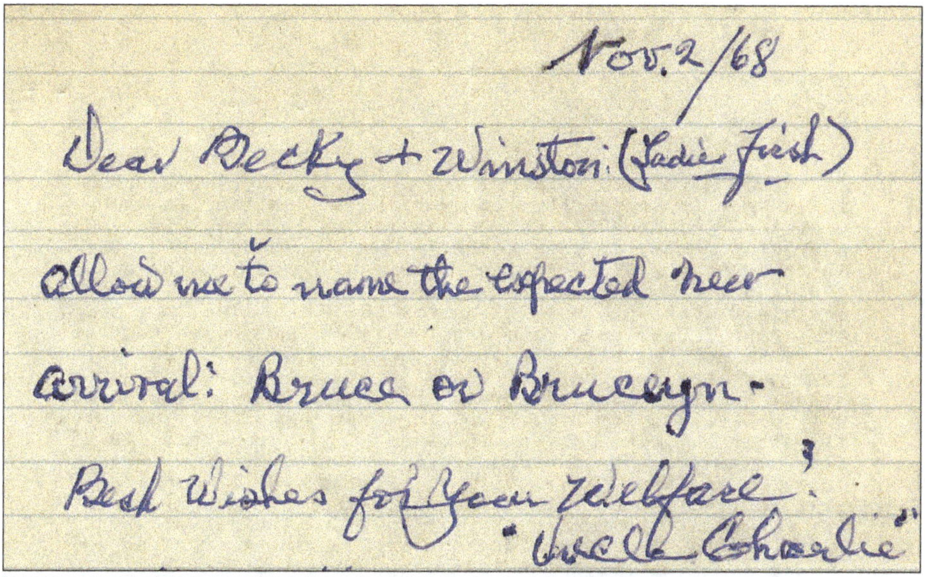

> Nov. 2/68
>
> Dear Becky + Winston (Sadie first)
>
> Allow me to name the expected new arrival: Bruce or Brucayn.
>
> Best Wishes for Your Welfare.
>
> "Uncle Charlie"

> OHIXIHO
> Feb 1/69
>
> Best Wishes to Andrea Jane and Parents.
>
> Enclose 2 smacks to be applied by Proxy.
>
> Trust all Well!
>
> "Uncle Charlie"

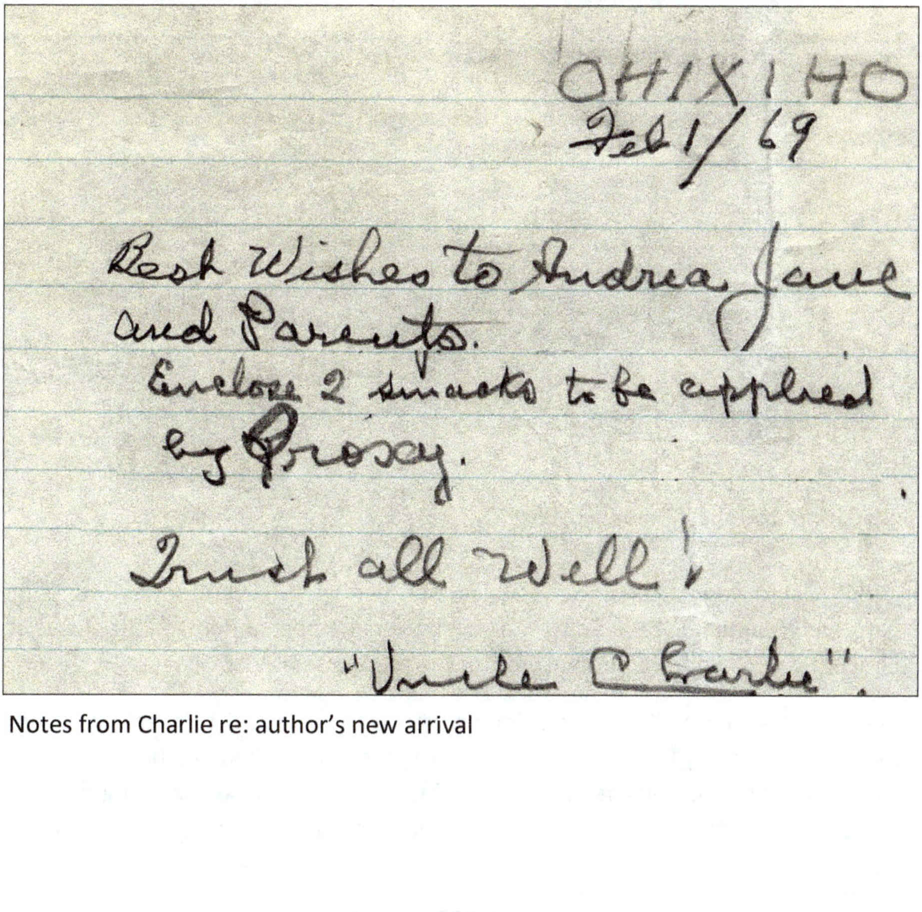

Notes from Charlie re: author's new arrival

> "OHIXIHO",
> Cookshire, Que., May 7/62.
>
> Miss. Mabel G. Fraser,
> 70 St. Usule,
> Quebec, Que.
>
> Dear Mabel:
>
> Through an acquaintance (cashier) at the Chateau Frontenac have just received your present address. Been wanting to get in touch with you ever since I noticed, in the Scotstown local of the Sherbrooke Record that you /the only surviving member of your father's family, since since the death of your sister, Dr. Ethel Fraser of Denver, Colorado.
> Well do I remember the times that you and your brother Gordon visited the old home place, "MAPLEMOUNT". Also, the time that your father made us a visit and how much he seemed to enjoy himself strolling through the maple woods. Do not remember ever meeting your older brother, ("Willie")? nor Ethel. My sister Nellie (Mrs. Planche) whenever in Quebec always called at your home. Looking over an album of snapshots the other day, came across a picture of you. Reminded me of how active you were in educational matters. The last twenty years or so, have passed very rapidly to me. I (Charlie) will be eighty-one, this coming June and am the only surviving member of the family of four boys and two girls: Jared, James, Nellie, Henry, Charles and Hattie,- in order of age.
> There are no surviving heirs except Gloria Fraser Bellam,- she has a family of three boys. Have just about completed a deal, disposing "MAPLEMOUNT" to a Welfare Organization for a home for boys. Since Nellie passed away in 1957, have been keeping house and living alone. Am very fortunate to be able to look after myself in every way,- still am quite active physically. One of my hobbies is coin collecting, chiefly U.S. and especially Old Canadians,- also, interested in curios.
> Am I correct in stating that your grandfather and mine were brothers ? If you have a biography or write up of Ethel's career, would enjoy perusing it and would return promptly.
> Would be very pleased to have a line from you Mabel, sometime when convenient and you feel inclined. Trust you have had a good Winter, free from colds and other ailments.
> The best I can wish for you, the best assett of all, HEALTH. If ever in Quebec, will surely call to see you. Typing is is easier on the eyes than my puzzling scrawl!
>
> My best respects to an old-time acquaintance, friend and distant COUSIN.
> CHARLIE
> C. C. Fraser.

Initial letter to cousin Mabel Fraser

accept such a verdict, Charlie massaged the man's legs several times a day for many weeks until he was able to walk again. Had he been a Roman Catholic, that miracle alone could have put him well on the road to sainthood!

Charlie was a man with strong opinions who was not afraid to speak his mind. Cousin June (Fraser) Patterson recalls an incident that happened when she was a very young child. "Once when Charlie and his Uncle John Rankin came for a visit, I offered Mr. Rankin a cookie. When Mr. Rankin said 'No, thank you,' Charlie

```
                                          "OHIXIHO"
                                Cookshire, Que., May 14/62.
Dear Mabel:
           Delighted to receive your gracious and  lovely
letter. I know that you will pardon the delay in answering!
  Recognize an old-time FRASER trait in your proffered hos-
pitality.  Appreciate very much your kind invitation which
pleased me very much. Later on in the Season expect to be in
your city on business for a couple of days.  Then, we will
have the opportunity to talk over old times et cetera!
  No! have not had a car the past few years, as I find  it
more convenient to travel by"Bus". Also, after a certain age
there are many restrictions for a Driver's License. Have
always been a great pedestrian, even now, a stroll of fifteen
or twenty miles does not bother me! Am a lover of nature,-
like to get out in the open spaces!  Please don't reprimand
me for talking too much about myself. Yes! we will without
any doubt find plenty to talk and discuss about when we meet.
  Now, about that "miserable"( wish could find a stronger word)
"i" in FRASER:- grandmother Frasier was from the U.S.A., She
thought FRASER sounded too Scotchy, so she wished to America-
nize it. She had the parson insert an "i" when  they  were
married. Imagine it,if you can? All the members of our fam-
ily were baptized "Frasier". Personally, have never used or
recognized the inserted "i". Nor did my brother Jared, the
rest, as baptized. Once when writing me, Jared commented,-
"Grandmother tried to make a bastard name out of the HONORED
NAME of FRASER. Was he angry!
  Gloria is well and lives in the home her father built,-about
one hundred yards from "MAPLEMOUNT".Her mother lives with the
family. Jim, met her in Florida; her home was in Evanston,Ill.
  Thank you for enclosing Ethel's Obituary which I am sending
herwith herewith, together with acopy which you may find use-
ful! Ethel must have been exceedingly clever and lived to a
good age to enjoy her wonderful career. would have liked to
have known her! How much you must miss her. 'Tis hard to have
to part with our loved ones!
                    Being in the habit of writing
business letters solely, I am venturing on new ground, if
Short Hand was within my ken, could say more with less words,
and thus relieve your eyes.  Very few letters have I enjoyed
asmuch as yours,Mabel! You write so free and easy,the words
seem to come right from your heart,as though I was in your
presence and you were speaking directly to me. Whats more!
You don't hesitate to meet one half-way,-or more!
  Good-bye for now.             Love and Best Wishes,
                                  From an old Cousin
```

Follow-up letter to cousin Mabel Fraser

reprimanded him very strongly, saying 'That is no way to teach a child; you should take the cookie and say thank you!' So Mr. Rankin accepted the cookie."

Niece Gloria's earliest memories of "Uncle Charlie" were when he would come to their house for Sunday afternoon visits. Her father, Jim Frasier (note the "i") and Charlie would always have very serious arguments about politics. "They would pound the table as they argued. At five o'clock Uncle Charlie would stand up, they would have a laugh and Uncle Charlie would go back home. The same routine

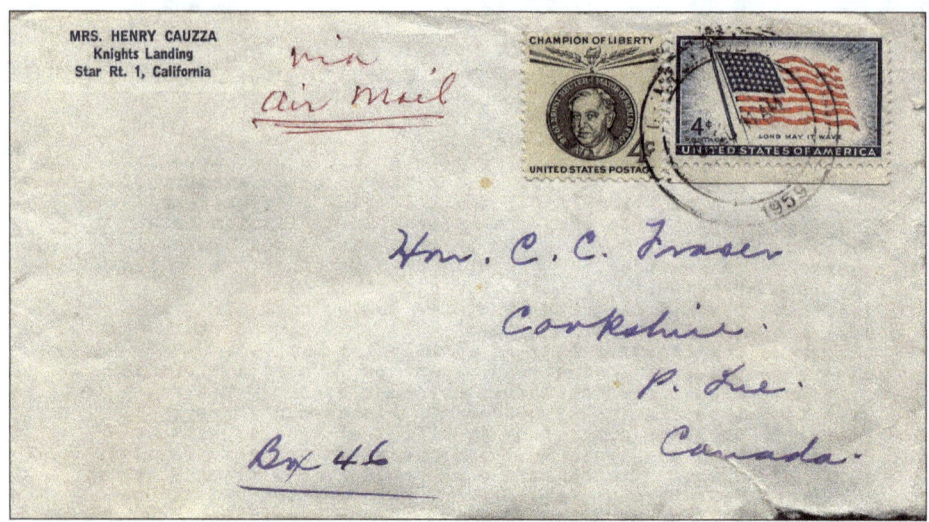

Cover to Hon. C. C. Fraser from his cousin Neva

would be repeated week after week." Two of Gloria's sons, Frasier and Liles, also remembered Charlie's Sunday visits: "He came to Sunday dinner on occasion, but we were at the age where 'children should be seen and not heard.' When the adults retired to the living room, we were sent outside to play."

Cousin June remembers that Charlie and his sister Nellie were very musical. "Nellie played the piano and everyone would sing. Charlie's favourite song was "After the Ball" (the biggest hit of the 1890s) and he would sing it to us often." The lyrics obviously resonated with Charlie the lifelong bachelor. In fact, Charlie wanted this song to be sung at his funeral – a request that was not honoured because it was felt to be inappropriate. The song was replaced by "God be with thee till we meet again," a hymn written by Charlie's ancestor, Rev. Jeremiah Rankin.

Charles Clark Fraser was a man of strong principles and unquestioned integrity. He successfully resisted two of the common vices of his day – drinking and smoking. He was no doubt influenced in this regard by his mother, Fanny Rankin. As a member of the Women's Christian Temperance Union, she signed a formal pledge in 1875 never to drink alcohol. And, according to Charlie's niece Gloria: "She made the family also sign this oath never to touch alcohol."

Charlie was never known to swear. Even if he would hit his thumb with a hammer, the strongest language he would use would be something like "Oy-joy!" His favourite expressions were "By Jove!" and "By eighty, Boy!" (or "By eighty, Girl!"). Due to his speech impediment, he pronounced the latter expression "By ay-ee, Boy!" (or "By ay-ee, Girl!").

In terms of religion, Charlie was of Congregationalist extraction, having been baptized in Eaton Congregationalist Church. But in his autobiographical summary, he specified his religion as "Libertarian: according to Declarer's revised definition; freedom from bigotry, etc."

Although not a regular churchgoer, Charlie was most certainly a man of faith. This is evidenced, among other things, by his support of the Canadian Bible Society and his interest in Billy Graham's evangelistic ministries. Among Charlie's papers was the following extract, entitled "Life a Continuous Film," from a Billy Graham pamphlet:

> Dr. Wilder Penfield, director of the Montreal Neurological Institute, said in a report to the Smithsonian Institute: 'Your brain contains a permanent record of your past that is like a single continuous strip of movie film complete with sound track, this film library records your whole waking life from childhood on. You can live again those scenes from your past, one at a time, when a surgeon applies a gentle electrical current to a certain point on the temporal cortex of your brain.' The report goes on to say that as you relive the scenes from

After the Ball

A little maiden climbed an old man's knee,
Begged for a story – "Do, Uncle, please.
Why are you single; why live alone?
Have you no babies; have you no home?"
"I had a sweetheart years, years ago;
Where she is now pet, you will soon know.
List to the story, I'll tell it all,
I believed her faithless after the ball."

Refrain:

> *After the ball is over,*
> *After the break of morn –*
> *After the dancers' leaving;*
> *After the stars are gone;*
> *Many a heart is aching,*
> *If you could read them all;*
> *Many the hopes that have vanished*
> *After the ball.*

Verse 2:

Bright lights were flashing in the grand ballroom,
Softly the music playing sweet tunes.
There came my sweetheart, my love, my own –
"I wish some water; leave me alone."
When I returned dear there stood a man,
Kissing my sweetheart as lovers can.
Down fell the glass pet, broken, that's all,
Just as my heart was after the ball.

Verse 3:

Long years have passed child, I've never wed.
True to my lost love though she is dead.
She tried to tell me, tried to explain;
I would not listen, pleadings were vain.
One day a letter came from that man,
He was her brother – the letter ran.
That's why I'm lonely, no home at all;
I broke her heart pet, after the ball.

After the Ball, by Charles. K. Harris

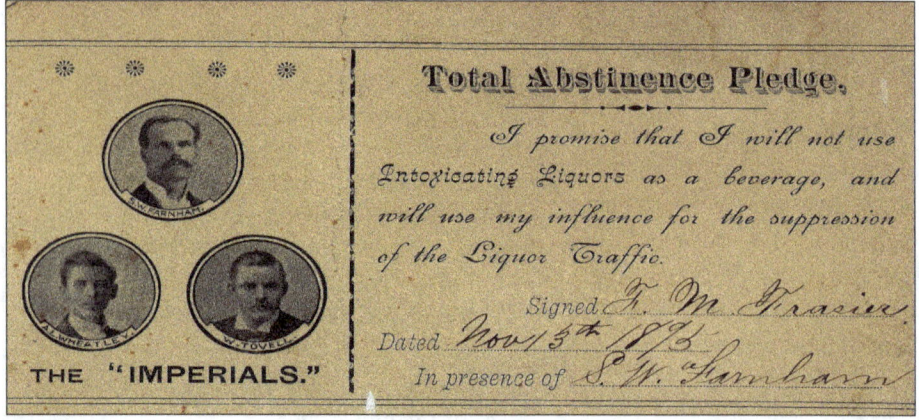

Total Abstinence Pledge signed by Charlie's mother (Courtesy of Gloria (Frasier) Bellam)

your past, you feel exactly the same emotions that you did during the original experience. Could it be that the human race will be confronted by this irrefutable record at the Judgment bar of God when He 'shall judge the secrets of men . . .' (Romans 2:16) ?

Charlie's gravestone contains a Bible verse that may have reflected the inner feelings of this intensely private person. The inscription "MENE MENE TEKEL UPHARSIN" from Daniel 5:25 has been interpreted in different ways. Known commonly as the "Writing on the Wall," the whole expression was interpreted as a prophecy and warning to the King of Babylon: "God hath numbered thy kingdom, and brought it to an end." However, portions of the writing have yielded their own interpretations. For example, the word TEKEL has been interpreted to mean "You have been weighed in the balance and found wanting." Is it possible that Charlie was expressing his own humility and insecurity? That will never be known because the entire content of his gravestone inscription was only revealed after his passing. Until then, his monument had contained only a puzzling partial inscription:

```
E E
    E E
        E E
            I
```

Charlie was a proud Canadian who openly displayed his patriotism. The large maple leaf flag on the sturdy wooden flagpole beside his house was very visible to all who passed. Beside it was an attractive stone monument that Charlie designed and installed in 1967 to commemorate Canada's centennial and Expo 67 (Montreal's World Fair). One of the lines engraved on it was the Latin expression "A POSSE AD ESSE" that means "from possibility to actuality."

Charlie's gravestone, Cookshire Cemetery (Photo by author)

In his autobiographical summary, Charlie stated that he had lived under every Canadian prime minister since Confederation except one. He noted that he remembered seeing Sir John Alexander Macdonald, the first prime minister of Canada. Charlie also declared that he had voted in every election ("dominion" and provincial) since 1904. As an unabashed Canadian nationalist, he detested the term "American" when referring to a citizen of the United States. "They are not

Americans; we are **all** Americans (i.e., North Americans); they are 'United Statesers'!"

Charlie was also very proud of his Scottish heritage. His paternal great-grandfather, Donald Fraser, emigrated from Scotland to Quebec in 1790 as part of the Highland Clearances. Together with three other members of his family, Donald made the dangerous two-month Atlantic crossing on the three-masted sailing ship "British Queen" during hurricane season. It is clear where Charlie inherited his penchant for adventure! In his home were various plaques and pictures that celebrated his Scottish background, including a Clan Fraser crest and a reproduction of "The Gathering of the Clans."

Centennial of Canada monument (Photo by Warren Fraser)

To say that Charlie was eccentric would be an enormous understatement. He was unconventional in so many ways. One example was the way that he dressed. Grandnephew Frasier Bellam remembers from his Sunday visits: "Uncle Charlie was always dressed in a three-piece suit, with a pocket watch and chain." But this was not only his Sunday apparel, it was his everyday dress as well. Whether digging potatoes in summer or cutting wood in winter, he could always be seen wearing a shirt, tie, vest, suitcoat and fedora. Cousin Charles W.K. Fraser recalls that sometimes Charlie would wear winter boots in the summertime.

Cousin Marilyn (Fraser) Reed remembers how fascinated Charlie was with dreams. "Whenever he would have a dream, he would force himself to wake up and write it down. Then he would go back to bed. The next morning, he would look up the dream in his dream interpretation book."

Some other things that he did were even more unusual. Once he made a bizarre request when he applied to the Court for permission to witness the hanging of a convicted murderer at Sherbrooke prison in 1931. It is unknown whether his request to observe the execution of Antonio Poliquin was granted.

Being a bachelor and living alone for a good part of his life, Charlie had to answer to nobody. He would disappear for weeks at a time without anyone knowing of his whereabouts. It could have been to attend a coin or stamp convention in Philadelphia, to visit cousins in California or to undergo surgery at the Royal Victoria Hospital in Montreal. He protected his personal privacy with passion.

Charlie the Man

Cousin Marilyn (Fraser) Reed remembers how Charlie mastered the art of changing the subject if he did not wish to answer certain questions. "That's a nice shirt you're wearing, Boy!" was enough to distract anyone whom he felt was trying to get too personal.

Although he was reserved, even shy, Charlie certainly had a sense of humour. Cousin Marilyn recalls how Charlie responded to itinerant Jehovah's Witnesses intrusions into his life. "See here, Girl (or Boy)! There are three things I do not discuss with anyone – politics, religion and my love affairs." This usually led to the polite but hasty retreat of the unwanted visitor(s).

Clan Fraser plaque (Photo by author)

Cousin Donald Parsons recounts an amusing story about Charlie that he had heard from his grandmother, Nellie Parsons, of Bury, Quebec. As Nellie's family would drive past Charlie's home (named OHIXIHO) in Cookshire, she would tell them: "He

Reproduction of "Gathering of the Clans" (From xmarksthescot.com)

has said that if any woman could tell him what OHIXIHO meant, he would marry her!"

Charlie also enjoyed playing the occasional trick on family or others. Cousin Charles W.K. Fraser tells about such an instance. "Once when Charlie was up North prospecting for an extended period of time, he grew a thick bushy beard. On the train home from Sherbrooke, he completely fooled his brother Jim who didn't recognize him."

When I was in high school, Charlie showed me a supposed Latin phrase that had been engraved on a desert hitching post and wondered whether the school principal might be able to interpret it. So, the next day I – the unsuspecting teenager – asked my principal what the "Latin" phrase TOTI EMUL ESTO meant. Without a moment of hesitation, the principal replied, "It means 'Be equal to the whole.'" I was too embarrassed to tell the principal the truth after Charlie revealed to me that it was not Latin at all, but stood for "TO TIE MULES TO"!

Epilogue

Charles Clark Fraser was truly one of a kind. He was a generous, humble and private person who loved family and accomplished so much in such an unusual variety of vocations. He was loved, respected and yes, intrigued, by all who knew him. His like will not be seen again. An indication of the esteem in which he was held is the fact that no less than three of his later generation cousins were named after him.

Charlie with namesake Charles Cameron Fraser (author's son), 1974 (Photo by author)

And now a personal word to Charlie himself:

As you enjoy a well-earned rest in your celestial home, may you rejoice in the knowledge that your life story has finally been told. And may you be happy with how it has been told. Thank you for the privilege of being your story-teller. Until we meet again, "Don't forget to exercise!"

Appendix A   Autobiographical Summary

Charles Clark Fraser's autobiographical summary, updated in 1973

OHIXIHO

These autobiographical notes were typed on Charlie's trusty old manual typewriter, an early 1900s vintage Remington Standard No. 7 model. Still in working order, the machine is now part of the Compton County Historical & Museum Society's collection at their museum in Eaton, Quebec.

Charlie's Remington Standard Typewriter No. 7 at Eaton Museum (Photo by author)

Appendix B   Fraser's Halo Complex of the Sun, March 21, 1930

FRASER'S HALO-COMPLEX OF THE SUN, (March 21, 1930).

Avery rare Halo-Complex of the Sun was observed by the author, C.C. Fraser, from the snowy frozen surface of Colnet Lake (Fraser's Lake) at the Canadian Gold Hunters' Mining Camps, Montbray Township, Abitibi County, Province of Quebec, Canada, (16 miles N.W. of Rouyn, P.Que.) on March 21st, 1930 at 2 hours 50 minutes P.M., Eastern Standard Time.

Approximate position at the time of display: Latitude N. 48.12, Longitude W. 79.22: Position of Sun, S. 60 W.: Observed duration, about minutes – 40: Temperature, 12 Fahn, – early in the morning, 6 degrees below zero.

The observer's location happened to be at a particularly favorable one, being at an elevation of some 900 feet. Lake Dasserat (913 ft.) and Lake Duparquet (882 ft.) are connected by the Kanasuta River into which the outlet of Colnet Lake (Fraser's Lake) flows. Both the River and Lake Dasserat are about three-quarters of a mile from Colnet Lake. Lake Dasserat has two outlets, one, the Kanasuta River, flows North through a string lakes and streams and ultimately reaches James Bay and The ARCTIC OCEAN. The other outlet flows southward, and in time reaches the head-waters of Upper Ottawa River and eventually to the ATLANTIC OCEAN. Therefore, Colnet Lake (Observer's location) is practically at the height of land.

The author's original sketches made on the spot, and diagrams and pertinent data, duly confirmed and attested by witnesses are the only extant, made of this very rare Halo-Complex of the Sun, at the time and place of the occurrence. Consider that this Halo compares quite favorably with the remarkable Haloes observed by Gassendi in 1630, and by Hevelius in 1661, and thereupon is worthy of note. We understand that alike atmospheric effects of such magnitude, grandeur and completeness may not occur again during the next 100 years or more, and be observed and depicted.

C.C. Fraser.
"OSIXIKO"
Cookshire,
P.Q.

N.B.
The original sketches were made hurriedly on the surface of the lake, in the heart of the bush. Probably, there will be scientific errors, as the drawings were made within a time limit with crude instruments. Prospectors camps are not usually equipped with the latest scientific facilities.
Possibly, some interesting features were omitted or unrecognized. From enquiry, no one in the surrounding area had ever seen/of the nature comparable in grandeur and completeness.   /ANYTHING/

Fraser's Halo Complex of the Sun, page 1 of 4

Page 2

FRASER'S HALO-COMPLEX OF THE SUN (March 21st.,1930).

CHARACTERISTICS:
Composed of four Circles and an Arc, Three Circles around the Sun, and one passing through it.

Apparent diameters of Circles A – B – C – D
48, 40, 28, 20 inches respectively and proportionally.
12, 10, 7, 5

Width of rings, ¼ of 1 inch

Arc "X" with 12 inch chord, ½ of 1 inch band /wide/

SUN'S apparent diameter, 1/8 of an inch

Above are all perspective measurements taken with a foot-rule at arm's length,- 18 inches from the normal eye.

Circle "A", the largest circle cut through the SUN

Circle "B", the next smaller circle, appeared to cut circle "A" through the center

Circle "C" was intersected at the S.W. circumference by the smallest Circle "D" by the width of the ring Circle"C"is especially interesting, being an irregular Halo Circle "D" had the SUN as center

ARC"X" appeared to be in the center of circle "A" on the circumstance of Circle "B"

The ring of the largest circle was of a whitish golden color. The rings of the other circles appeared to be tinged with reddish orange. The brightest spots were at the junction points of the largest and smallest circles. The ARC "Y" from the South junction of the Circles "A" and "B" to the horizon was prismatically colored.
The ARC "X" was not observed until towards the close of the display, and was of a very deep reddish color, and was the last to fade out.
The center of the largest Circle "A" appeared to be directly overhead of the observer: the other three circles were South 60 W. from the observer. All the circles were on the same axis.
The Eastern horizon was practically lake-level forest. South and N.W. range of forest covered hills, 30 to 60 feet in height.
SUN appeared to be in a slight haze at peak of display,- could almost look at the SUN directly with the naked eye, Sun growing brighter as the display faded.

In the sketches or diagrams the size of the SUN, also the widths of the rings, Arc and bands have been increased disproportionally relative to the sizes of the circles, in order to display to advantage.

C.O.Fraser
"OHIXIHO"
Cookshire
P.Q.

Fraser's Halo Complex of the Sun, page 2 of 4

Fraser's Halo Complex of the Sun – March 21, 1930

FRASER'S HALO-COMPLEX OF THE SUN (March 21st., 1930).

Position of SUN at occurrence of display:- S. 60 W.

Observed duration: forty minutes, 2.50 P.M. - 3.30 P.M., E.S.Time
Appeared to be at peak of display when first observed.
TIME:-
    March 21st., 1930, 2.50 P.M., EASTERN STANDARD TIME
    March 21st. 1930 Vernal Equinox
    March 21st., 1930 Last Quarter of the Moon .

WEATHER Chart:-
    Temperature 12 degrees Fahrenheit
    Elevation 900 feet

March 21, 1930: Early in the morning, 6 degrees below zero. The day was clear and bright with N.E. wind of about 20 miles; a few very light scattered clouds: SUN set in light clouds; night, clear and starry. Preceeding few days, light snow-falls and windy. TEMPERATURE:- from freezing to zero.

March 22nd. Bright and fine, fleecy clouds; freezing to zero: night clear and bright.
March 23rd. Fine and clear all day, Sun set in clouds. Night, starry and clear; temperature;- zero to freezing.
March 24th. Morning, clear and bright; few light clouds. Temperature: 10 Fahn. to freezing; night, clear.
March 25th. similiar to 24th., except, getting cloudy towards evening, turning to wind and snow; snowed all night.
March 26th. still snowing, very heavy snow-storm and wind during the night; snow-fall of sixteen (16) inches.
March 27th. 28th. and 29th. fairly mild and cloudy; TEMP. about freezing.

Coincidently, the evening of the same day of FRASER'S HALO-COMPLEX OF THE SUN, there occurred at Colnet Lake at 8.50 P.M., E.S.Time, an arched BAND," Z ", at the N.E. horizon, with a CHORD of 36 inches; width of band ½ inch, very deeply colored. Gradually getting wider as it faded, thence sending out streamers over the Northern sky. A most magnifixient display of AURORA BOREALIS. It was first observed at its peak.
The center of the Arc appeared to be at a point where the N.E. circumference of CIRCLE "A" had been at the time of the phenomenon of the Sun, and on the same axis.

                                          C.C.Fraser
                                          "OHIXIHO"
                                          Cookshire
                                          P.Q.

Fraser's Halo Complex of the Sun, page 3 of 4

RE – FRASER'S HALO-COMPLEX OF THE SUN (March 21st.,1930).

Extract from the R.A.S.C. JOURNAL of 1930, Volume XXIV, Page 237:-
A SOLAR HALO SEEN AT HAILEYBURY, ONT.

Mr W.H.Luke, of Haileybury,Ont. sends the sketch reproduced herewith of a halo seen on MARCH 21st. from 4 P.M. until sundown. The mock suns on either side of the Sun "S" were at times too brilliant to look at with unprotected eyes. The outer Circle though incomplete and not so brilliant as the inner one, was very beautifully colored, especially at the part where it joined the reverse Arc, At times it would compare with a Summer rainbow.

////////////////////////

THE ABOVE DESCRIBED HALO SEEN THE SAME AFTERNOON AS "FRASER'S HALO!. --
-WHICH WAS OBSERVED ON COLNET LAKE.
Haileybury,Ont. would estimate,is some twenty miles S.W.   C.C.Fraser
                                                          "OHIXIHO"
                                                          Cookshire
                                                          P.Q.

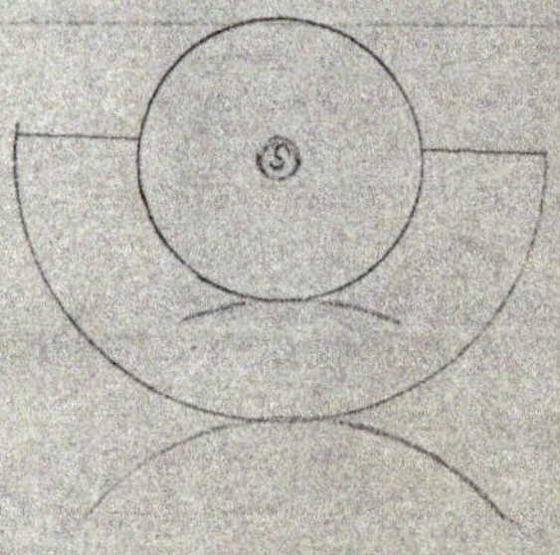

Luke's Halo

Fraser's Halo Complex of the Sun, page 4 of 4

Appendix C   Birth/Baptism Record, Death Notice and Obituary

Charles Clark Fraser record of birth and baptism (Digitally enhanced by Greg Beck)

FRASER, Charles Clark — At the home of Mrs. Katie Burton, Lennoxville, Que., on Wed. Oct. 25, 1978, Charles Clark Fraser of Cookshire, in his 98th year, the last of the family of James Augustus Fraser and his wife Fanny Maria Rankin. Resting at the Gordon Smith Funeral Home, 120 Main St. West, Cookshire, where funeral service will be held on Sat. Oct. 28 at 3 p.m. Interment in the family plot in Cookshire Protestant Cemetery. In lieu of flowers, donations to the Canadian Bible Society, Montreal, would be gratefully acknowledged. Visitation on Thurs. from 7 to 9, Friday 2 to 4 and 7 to 9.

Above: Death notice
Opposite: Obituary
(Sherbrooke Record)

THE SHERBROOKE RECORD — FRI., DEC. 1, 1978

Obituary

CHARLES C. FRASER
of Cookshire

Charles Clark Fraser, a life-long resident and one of the oldest living natives of Cookshire, passed away on October 25, 1978, in his 98th year.

He was born in Cookshire on June 29, 1881, the son of James Augustus Fraser and Fanny Maria Rankin. He received his formal education at the Cookshire Academy.

Mr. Fraser resided most of his life at the Fraser family home "Maplemount" before moving to "OHIXIHO" in 1958. Six years ago, he moved to Lennoxville where Mrs. Katie Burton, a long-time family friend, provided him with a happy home and excellent care for his remaining years.

"Uncle Charlie", as he was fondly known to many, was loved and respected by all who had the pleasure of knowing him. To organizations and individuals alike he lived the motto that "a friend in need is a friend indeed". He was a director of the Maplemount Home and was a life member of the Canadian Bible Society and of the Cookshire Christmas Club.

His wide range of interests and his boundless ambition were reflected in the many occupations which he held during his working life. He was a chemist at Frasier, Thornton & Co., a mining prospector and developer in Northern Canada, a silver fox rancher, a farmer, a trapper, a woodsman, a mechanic, a pipe-fitter, a stationary steam engineer and a confectionary manufacturer.

Charlie Fraser was a strong advocate and lifelong practitioner of physical fitness. Although of slight build, he was a man of abnormal strength and endurance. Among his unusual physical accomplishments were: lifting 650 lbs. with one hand, running 100 yards in 10 seconds, walking 65 miles in one stretch, and extracting his own infected molar with a pair of pliers! Besides exercising, running, walking, and cycling, he also played several sports including cricket, baseball, hockey, and lacrosse. He enjoyed hunting and fishing.

His hobbies included radio experimentation, astronomy (he observed and documented a spectacular halo-complex of the sun), coin and stamp collecting, mathematics, history, and word-coining. He was a life member of the Royal Astronomical Society of Canada and of the Pioneer Philatelic Phalanx and a member of the Canadian Numismatic Association. He was also a life member of the Compton County Historical & Museum Society and was curator and founder of the OHIXIHO Museum and the Ancient Order of Petrologists.

He is survived by his niece, Gloria (Frasier) Bellam of Cookshire, three grand-nephews, and many cousins. He was predeceased by his seven brothers and sisters — Bailey, Lily, Jared, James, Nellie, Henry and Hattie.

The funeral was held on October 28, 1978, at the Gordon Smith Funeral Home in Cookshire. A large number of friends and relatives overflowed the chapel. The service opened with a favorite hymn of the deceased - "Abide With Me". Dr. W.J. Klinck, of Lennoxville, a close friend and associate, conducted the service. "God Be With You Till We Meet Again", a hymn composed by Dr. J.E. Rankin, an ancestor of the deceased, was played by Mrs. Gordon Patterson.

The bearers were Harold Bellam, Malcolm Fraser, Pierre Beaulieu, Paul St. Laurent, Bryan McDermott and Ray Stevenson. Interment was in the family plot at the Cookshire Cemetery with Rev. Terry Blizzard officiating.

Charles Clark Fraser will be greatly missed by all who knew and loved him.

Appendix D Frasier, Thornton & Company Raw Materials Price List

WHITE PINE BARK NO. 12 GRAN.,	.29	per lb
SPIKENARD ROOT NO. 12 GRAN.,	.60	" "
SENEGA ROOT NO. 12 GRAN.,	2.10	" "
BLOOD ROOT NO. 12 GRAN.,	.65	" "
LICORICE ROOT NO. 12 GRAN.,	.40	" "
GRANULATED SUGAR	7.80	per 100 lbs
WILD CHERRY BARK	.17½	" lb
HENBANE LEAVES	.60	" lb
SASSAFRAS BARK	.69	" lb
SQUILL ROOT	.50	" lb
ANTIMONY POT TART	.25	Oz
ALCOHOL	4.00	per Gallon
GLYCERINE	.38	per lb
CHLOROFORM	.48	" "
BURNT SUGAR	1.70	" gallon
AMBER PETROLATUM	8.25	" 100 lbs Collect
PARRAFIN WAX	.15	" lb
CAMPHOR	.66	" "
CARBOLIC ACID CRYSTALS	.65	" "
OIL SASSAFRAS	2.50	" "
BEESWAX	.50	" "
RESIN	.10	" "
LARD	.25	" "
SALICYLIC ACID	1.24	" "
TALC	17.50	" Ton

Raw materials price list, page 1

BORIC ACID POWDER	.12	"	lb
PERFUME	4.00	"	Gall(Muskalene)
JAMAICA GINGER ROOT	.73	"	lb
OIL LEMON	5.25	"	lb
OIL OF CEDAR LEAF	2.00	"	lb
OIL EUCALYPTUS	1.05	"	lb prepaid
CAMPHOR POWDERED	.66	"	lb
MENTHOL	10.10	"	"
WHITE PETROLATUM (.No.2 Pain Ease Rub.)	14.20	"	100 lbs
PARRAFIN WAX	.15	"	lb
IMPORTED CASTILE SOAP	.21	"	lb
GUM SPTS. TURPENTINE	.89	"	Gallon(collect)
SAL AMMONIAC WHITE POWD. MEDICINAL	.10	"	lb
LIQUID AMMONIA FORT 880	.11	per lb	
SACCHARINE	3.20	" "	(1947)
OIL PEPPERMINT	8.50	"	lb
OIL BITTER ALMONDS	6.50	"	lb
OIL CLOVES	5.75	"	lb
GROUND SULPHUR	4.20	per 100 lbs	
SABADILLA SEED	.45	"	lb
CHARCOAL	110.00	"	Ton
DARK VETERINARY GREASE	Not used now		
AMBER PETROLATUM GREASE	8.25	per 100 lbs (coll	
OIL OF PINE TAR	.80	"	Gall
MERCURY PERCHLORIDE U.S.P. POWDER	1.40	"	lb
MENTHYLATED SPIRITS (.we use now Stansol 95%)	.93	"	Gallon
THICK OIL OF PINE TAR	.80	"	"
TURPENTINE	.89	"	"
SALTPETRE PURE (NITRATE OF POTASH)	.14	"	lb
SULPHATE IRON (SUGGAR COPPERAS)	4.10	"	100 lbs
SHORTS	2.65	"	"
RED OXIDE	9.90	"	"
PURE POWDERED ALUM (POTASH)	5.00	"	"
SODIUM SULPHATE ANHYDROUS	4.25	"	"
GINGER	.39	"	lb
MAGNESIA CARBONATE LIGHT	10.75	"	100 lbs
POWDERED SOAP	.75	"	lb
CAPE ALOES	.56	"	lb

Raw materials price list, pages 2-3

Charles Clark Fraser

"A well-written biography of Charlie, a highly intelligent, fascinating and most unusual man who deserved to have a book written about his life."
– Shirley Nadeau, copy editor of the *Quebec Chronicle-Telegraph*

"Charles Clark Fraser was a remarkable man, and this is a remarkable book. It was carefully researched and the visual material presented makes it stand out from a normal work of family history."
– Roland Kuhn, author of *Apocalypse North: Who Killed Stebben Harpoon?*

"In this book Winston Fraser has honoured a forebear's wish by taking the memorabilia of a life well lived to create a wonderful story. In an increasingly global world it is important from time to time to focus on the microcosm of our families as a way of grounding ourselves. Winston has done just that through telling Charlie's fascinating story."
– Dr. Sandy Fraser, retired professor of Education, Acadia University

A computer consultant by profession, Winston Fraser is a widely published photographer and writer. A book of his photographs, *Historic Sites of Canada*, was published in 1991. Fraser was the major supplier of photos for National Geographic's *Canada Travel Guide*. Most recently he has self-published *Endangered Species of Country Life*. He is a first cousin, once removed, of Charles Clark Fraser.

Front cover photo:
Courtesy of Charles W. K. Fraser
Back cover photo:
By the author, 1968

ISBN 978-0-9950842-0-9

9 780995 084209

Frasier, Thornton & Company Raw Materials Price List

RED COLORING (Red Oxide)	9.90	"	100 lbs
WILD STRAWBERRY LEAVES (COARSE GROUND)	.95	"	lb
GROUND LOGWOOD CHIPS	.65	"	lb
WILD BLACKBERRY ROOTS	.40	"	lb
RHUBARB ROOT	.75	"	lb
CREOSOTE (BEACHWOOD) B.P. WHITE	2.21	"	lb
OIL OF WINTERGREEN	.45	"	Oz
OIL OF CASSIA	.70	"	Oz
RAW LINSEED OIL	2.00	"	Gallon
KEROSENE	.269	"	" deliver
OIL MYRBANE	.25	"	lb

WILD CHERRY BARK NO. 12	.17½	per lb	
HYPOPHOSPHITES SODA	1.07	"	lb
MALT EXTRACT	.14½	"	lb
OL. GADUOL)			
MORRHUOL) Gadex	5.00	per lb	
JECORROL)			
OL. ANISE	3.00	"	lb
QUININE SULPHATE	.50	per Oz	
HYDROCHLORIDE ACID	.74	"	lb
WILD CHERRY EXTRACT	3.15	"	Gallon
SNOW WHITE PETROLATUM ... No. 2 (for Pain Pase.)	14.20	"	100 lbs
CREAM WHITE PETROLATUM	Not used		
GUM CAMPHOR	.66 per lb		
MENTHOL ...(Crystals Chinese U.S.P.)	10.10	per lb	
THYMOL	5.40	"	lb
WHITE WAX	.15	"	lb
SALICYLIC ACID	1.24	"	lb
COLOR VERA GREEN ..(Chlorophyll Soluble in Oil)	5.75	"	lb
COTTON SEED .... White Oil)	.23 per lb		
PETROLATUM OIL ..(Baby Oil. Risella. No. 17	.855	"	Gallon (Prepaid
DARK VETERINARY PETROLEUM	Not used now		
CAYENNE PEPPER	.77 per lb		
OIL CAJEPUT	4.00	"	lb
CANADA BALSAM	4.50	"	lb

Raw materials price list, pages 4-5

BALS TOLU	7.10	" lb
OIL OF PINE TAR HEAVY	.80	" Gallon
FLUID EXTRACT SQUILLS	Make this	
SYRUP WHITE PINE COMPOUND "PINEALCO"	2.15	per lb
SOLID EXTRACT (Worm Syrup)	5.40	" lb
CARMINE	8.50	" lb
SWEET PEA PERFUME	5.00	" lb
GARDENIA COMPOUND	10.00	" lb
OIL OF TAR MEDIUM WEIGHT PINE TAR	.80	" Gallon
COD OIL	.95	" "
ISOPROPYL	.93½	" " Prepaid
POT. BICARB	.87	" lb
SODA BICARB	4.75	per 100 lbs

CAPE ALOES POWDERED	.56	per lb
OIL JUNIPER BERRIES	1.00	per Oz
CASCARA BARK COARSE GROUND	.62	per lb
AFRICAN GINGER " "	.39	" lb
BURDOCK ROOT " "	.85	" lb
CARRAWAY SEED " "	.71	" lb
RHUBARB COARSE GROUND	.75	" lb
GENTIAN ROOT COARSE GROUND	.60	" lb
ORANGE PEEL BITTER COARSE GROUND	.32	" lb
OIL BERGAMONT	.25	" Oz
OIL CLOVES	5.75	" lb
LILACIN	.05¼	" Oz
VANILLIN	.35	Oz
CHLOROFORM	.48	per lb
OIL OF PINE TAR DARK	.80	Per gallon
CAYENNE PEPPER PODS (SMALL AFRICAN)	.46	per lb
GUM MYRRH	.90	per lb
OIL SPEARMINT 2	6.50	" lb
OIL THYME	2.60	" lb
OIL ROSEMARY	1.80	" lb
OIL LAVENDER SPIKE	2.60	" lb
PINE PITCH KIDNEY PILLS	2.20	" M
OIL THYME BPRED (RED)	2.60	per lb
OIL SASSAFRAS TRUE B.P.	2.50	per lb

Raw materials price list, pages 6-7

Frasier, Thornton & Company Raw Materials Price List

```
OIL CAJEPUT B.P. .......................................   .32½  "  oz
COTTON SEED OIL .......................................   .23   "  lb
CAMPHOR FLOWERS .......................................   .66   "  lb
                    2232
IOSOL 2232 ...(Excelsior Lighter Fluid.)............   .50 per gallon
BLACK COAL TAR DISINFECTANT ........................   .71 per gallon collect
ETHYL NITRITE CONC. ..................................   2.53  "  lb
VARSOL 3139 .(Spot Out)..............................   .45 per gallon
SQUILL ROOT NO. 12 ...................................   .50 per lb
Oil Mustard                                              .20 . oz.
Red Rat Squell                                           .70 . lb.
Orris Root                                               .60 . lb.
Rose Perfume                                             .28 lb.
Castor Oil                                               2.80 gal.
```

Raw materials price list, page 8

www.ingramcontent.com/pod-product-compliance
Lightning Source LLC
Chambersburg PA
CBHW070607010526
44118CB00012B/1465